NOBEL CONFERENCE XXVI

CHAOS:
The New Science

Ilya Prigogine, Nobel laureate, University of Texas and Solway Institutes
Mitchell Feigenbaum, Rockefeller University
James Gleick, author of *Chaos: Making a New Science*
John Holte, Gustavus Adolphus College
Benoit Mandelbrot, IBM Thomas J. Watson Research Center
Heinz-Otto Peitgen, University of Bremen
John Polkinghorne, Queen's College, Cambridge
Steve Smale, University of California–Berkeley

Edited by
John Holte

Gustavus Adolphus College
Saint Peter, Minnesota 56082

Copyright © 1993 by
Gustavus Adolphus College

University Press of America®, Inc.
4720 Boston Way
Lanham, Maryland 20706

3 Henrietta Street
London WC2E 8LU England

All rights reserved
Printed in the United States of America
British Cataloging in Publication Information Available

Library of Congress Cataloging-in-Publication Data

Nobel Conference (26th : 1990 : Gustavus Adolphus College)
Chaos : the new science / Ilya Prigogine . . . [et al.] ; John Holte,
    editor ; Nobel Conference XXVI.
        p.   cm.
Conference held at Gustavus Adolphus College in 1990.
1. Chaotic behavior in systems—Congresses.   I. Prigogine, I. (Ilya)
        II. Holte, John.   III. Title.
Q172.5.C45N6   1990   003'.7—dc20   92-34498 CIP

ISBN 0-8191-8933-2 (cloth : alk. paper)
ISBN 0-8191-8934-0 (pbk. : alk. paper)

The paper used in this publication meets the minimum requirements of American National Standard for Information Sciences—Permanence of Paper for Printed Library Materials, ANSI Z39.48-1984.

Contents

Acknowledgments — v

Introduction — vii
John Holte

Contributors — xiii

Fractals — 1
Benoit Mandelbrot

Discussion

The Causality Principle, Deterministic Laws and Chaos — 35
Heinz-Otto Peitgen

The Transition to Chaos — 45
Mitchell Feigenbaum

**Time, Dynamics and Chaos:
Integrating Poincaré's "Non-Integrable Systems"** — 55
Ilya Prigogine

Discussion

What Is Chaos? — 89
Steve Smale

Discussion

Chaos and Cosmos: A Theological Approach — 105
John Polkinghorne

Chaos and Beyond — 119
James Gleick

Acknowledgments

In the beginning Elvee said, "Do it," and there was Chaos.

The "beginning" for Nobel Conference XXVI came in 1988, when Richard Elvee, the Director of the Nobel Conference series, told me to go ahead and organize the conference we called *Chaos: The New Science*. The subsequent intellectual direction of the conference was guided by a committee of Gustavus Adolphus College faculty: C. Milton Brostrom, Richard Elvee, Thomas Gover, William Heidcamp, Michael Hvidsten, Michael Moon, Charles Niederriter, Brian O'Brien, Lawrence Potts, Ronald Rietz, and Jeffrey Rosoff.

The successful preparation and organization of the conference entailed substantial efforts on the part of many Gustavus staff, faculty, and students. We are especially grateful for the work of Elaine Brostrom and the Office of Public Affairs, who managed to turn our chaos conference into an orderly affair; Patrick Francek and the Office of Media Services, who met the challenge of putting on a major multimedia event; Linda Fullerton and the Food Service, who fed the crowds that more than doubled the on-campus population; Laura Stone and the members of the Art Department who arranged for the Peitgen exhibit, "Frontiers of Chaos"; and the music Department, which provided special music at all the lectures and at the Nobel Concert. Thanks are also due to the hosts who accompanied our speakers: Dr. David Seely and student hosts Nathan Blair, Jason Douma, Scott Hess, Michelle Hines, Lori Munter, David Sukow, and Dan Woodside.

The graphic design on the cover of this book and on the conference poster and brochure is a close-up view of the Mandelbrot set generated by Gustavus student Scott Hess. The posters and brochures themselves were produced by Kelvin Miller of Primarius, Ltd. For extensive secretarial services involved in preparing this manuscript for publication I thank Janine Genelin, Colleen Overson, Helen Delahunt, and Dawn Kamadulski.

Funding for the Nobel Conference is made possible by an endowment established by the late Russell and Rhoda Lund.

Finally, our most effusive thanks go to the seven principal contributors to this conference. In addition to their lectures, which we publish here, they freely participated in the Nobel "firing lines" and numerous other discussions, making our Nobel Conference the intellectually stimulating event that it was.

INTRODUCTION

John Holte

In recent years the images and ideas of a strange new science have been percolating into the public consciousness. The movie-going public has witnessed wondrous worlds of artificial reality swirl before its eyes, as in the *Genesis* scene in *Star Trek II: The Wrath of Khan*. Art lovers from the Guggenheim Museum in New York to the University of Santa Cruz have been spellbound by shows by Benoit Mandelbrot and Heinz-Otto Peitgen featuring the richly textured images of fractional dimension known as fractals. Computer hackers have become enthralled with the algorithmically generated pictures of a previously unseen world of mathematically entrancing patterns born of chaotic dynamics and fractal geometry. The educated reader has been offered glimpses of a revolutionary new science in a series of popular books: *The Fractal Geometry of Nature* by Mandelbrot, *Order Out of Chaos* by Ilya Prigogine and Isabelle Stengers, *The Beauty of Fractals* by Peitgen and P. H. Richter, and *Chaos: Making a New Science* by James Gleick. Because of the public fascination with the ideas of chaos and fractals, it was decided that the twenty-sixth annual Nobel Conference at Gustavus Adolphus college should be *Chaos: The New Science*.

It was our dream to have as speakers at the conference the brightest stars of the new science of chaos. We drew up our list of possible people to invite, and at the top of the list we had: Mitchell Feigenbaum, James Gleick, Benoit Mandelbrot, Heinz-Otto Peitgen, John Polkinghorne, Ilya Prigogine, and Steve Smale. We were immensely delighted when they all accepted our invitation. Evidently the public, too, was delighted to have the opportunity to learn from this distinguished panel, for on October 2-3, 1990, we had between 4,000 and 5,000 people in attendance, including representatives of 200 high schools and 140 colleges and universities.

What is chaos? There is disagreement among scientists on how it should be defined. Many would consider the hallmark of chaos to be sensitive dependence on initial conditions, also named the Butterfly Effect after Edward Lorenz's paper, "Predictability: Does the flap of a butterfly's wings in Brazil set off a tornado in Texas?" The term "chaos" itself appears to have been coined by Li and Yorke in 1975 to describe a kind of dynamical behavior more complicated than the familiar steady state or cyclic patterns. In 1986 an international conference convened by the Royal Society in London formulated the following definition:

"3. (Math.) Stochastic behaviour occurring in a deterministic system." This definition seems to be an oxymoron, combining randomness and determinism, and therein lies part of the fascination of chaos. Furthermore, as Prigogine notes in these pages, the connotation of "chaos" is too negative; chaotic dynamics can in fact describe stable or self-organizing systems.

The origins of chaos long predate these definitions. Its role in dynamical systems was glimpsed a century ago by the great French scientist Henri Poincaré, who, when he looked at the problem of the stability of the solar system, found the motions of three or more bodies under the influence of gravity to be so complicated that he recoiled in horror. An archetypal form of chaos—an "icon" of chaos, according to Feigenbaum—is turbulent flow in a fluid. A half century ago the great Russian mathematician A. N. Kolmogorov took on the challenge of understanding turbulence, and subsequently, together with Arnold and Moser, developed the KAM theory of chaos, as Prigogine notes in his lecture.

But the revolution in chaos theory is really a development of the recent past. Just three decades ago, Smale identified the crucial feature of dynamical systems that leads to chaotic behavior; he explains this in detail in his lecture for this conference. In the 1970s Feigenbaum discovered an astonishing regularity underlying chaotic behavior, the "universality" in the pattern of transition to chaos, and he speaks on this transition to chaos in his lecture. Meanwhile, Prigogine was pioneering a penetrating scientific and philosophical study of how order can emerge from chaos, a study that he presses forward today. The new study of chaos is so young that only recently have philosophers and theologians, including physicist-turned-theologian Polkinghorne, begun to grapple with the issues raised by chaos.

The dramatic advances in chaos theory made in the past three decades are superbly narrated in Gleick's *Chaos*. He chronicles the stories of many natural and social scientists who have been troubled by the inadequacy of the traditional "linear" view of the world, according to which changes in the outputs are proportional to the changes in the inputs. So they have turned to nonlinear models. As they have applied nonlinear models in their struggle to understand complex and irregular phenomena in a wide variety of fields, they have discovered an underlying simplicity and regularity. They have found that complex, unpredictable phenomena may have elegantly simple, deterministic models, and conversely, that simple, deterministic models may exhibit startlingly complex and unpredictable behavior.

The range of applications of chaos theory is awesome: information processing in the brain, protein formation, cardiac arrhythmias, epidemics, fluctuations in wildlife populations, business cycles, the arms race, coherent chemical oscillations, the configuration of river systems, fluid turbulence, the Great Red Spot of Jupiter, supergalactic structures, and even mathematics itself. The new science of chaos transcends the traditional divisions of science so that scientists describing phenomena in different disciplines now find that, deep down, they are talking about the same thing. The reason they can do so is that the deep-down essence of chaos theory, the common language that describes all these phenomena, is mathematics.

How do fractals relate to chaos? It turns out that fractal geometry provides the appropriate mathematical language for describing chaotic systems. It is not the geometry of straight lines and circles and spheres and cones, but the geometry of lightning bolts and coastlines and clouds and mountains. This geometry aptly describes intricate shapes, both natural and artificial, that seem irregular and chaotic, but exhibit a self-similar appearance at all scales of magnification. Fractal geometry, like chaos theory, has antecedents in the late nineteenth century, but unlike chaos theory, its foundation was primarily the work of one person: Benoit Mandelbrot. Following the trail he blazed, other researchers and popularizers, including Heinz-Otto Peitgen, have contributed to its glorious successes.

In the lectures collected in this volume, the level of presentation makes varied demands on the reader. But even when it is most technically demanding, if one pushes on, one will find that the lecturer brings the discussion back to earth, rewarding patience and persistence with new insight.

Benoit Mandelbrot has written, "clouds are not spheres, mountains are not cones, coastlines are not circles, and bark is not smooth, nor does lightning travel in a straight line." Thus the traditional geometry of Euclid offers caricatures of the objects of nature, but the new geometry of fractals can render uncanny likenesses of the features of our world. In his lecture, the founder of fractal geometry emphasizes that geometry, to be appreciated properly, must be presented with a strong visual component. Although we cannot present in book form the spectacular visual imagery he (and other lecturers) presented at the conference itself, we do have, through the courtesy of Dr. Mandelbrot, a nice set of figures for his illustrated guide, leading us through basic fractal geometry to the frontiers of research. His primer clearly explains the notions of fractal and fractal dimension in a down-to-earth fashion ("geometry," after all, means "earth measurement"). In addition, the discoverer himself provides a description of the now-famous Mandelbrot set.

Heinz-Otto Peitgen, whose portraits of fractals and chaos in books, magazines, and traveling art shows have enthralled millions of people throughout the world, captivated our Nobel Conference audience with a dazzling multimedia demonstration of the ways of chaos. We are unable to reproduce in book form his live computer experiments, his three-magnet demonstration, his supercomputer-generated video, and so on. So we are grateful to Professor Peitgen for providing us with a manuscript that focuses on fundamental philosophical issues surrounding chaos theory and puts them in a historical perspective. He argues that the Butterfly Effect of chaos theory and the uncertainty principle of quantum mechanics combine in a pincer movement attacking the principle of cause and effect.

Mitchell Feigenbaum, famous for his discovery of a wholly unexpected "universal" pattern in the progression toward chaotic dynamics, asks us to visualize a fluid in turbulent motion, in which streamlines break up into whirls and eddies at innumerable scales, as an "icon" of chaos. As an icon it provides us with a mental handle on the phenomena we are trying to grasp and suggests the sort of questions we would like to answer. From this starting point, he leads us through the modes of description that scientists employ in trying to comprehend chaos, from statistical distributions to geometry. After contrasting linear and nonlinear geometry, he explains—rendering the mathematics involved entirely into words—how it happens that so many strongly nonlinear phenomena can share a quantitatively identical geometry, i.e., how there can be universality in the transition to chaos.

Ilya Prigogine, who was awarded the Nobel Prize in chemistry in 1977 for his contributions to understanding non-equilibrium thermodynamics, also has a reputation for profound reflection on fundamental issues in science. For him, "chaos" is a misleading term because it emphasizes disorderliness and unpredictability and destructiveness when, in fact, so-called chaos can also be the basis for order and stability and structure, including life itself. In his lecture (and the subsequent discussion) he marshals the insights of recent research in dynamics to address the question, What is time? In particular, how can we resolve the paradox of the "arrow of time"? (Although we experience time as flowing from past to future, this experience, as Einstein said, is an "illusion" not grounded in the laws of physics.) Unwilling to accept a formulation of the laws of science contradicted by experience, Prigogine finds that the time one experiences is naturally associated with the building up of correlations. The fact that these correlations are not an illusion is a consequence of a fundamental theorem of Poincaré, which Prigogine ranks among the most significant discoveries of science. Prigogine then introduces us to Large Poincaré Systems and presents for us here some of the technical details of his pathbreaking research into this more inclusive form of dynamics.

Steve Smale, who in 1966 was awarded the Fields Medal (generally regarded as the equivalent of the Nobel Prize for mathematics), likewise carried forward the legacy of Poincaré with his solution of the long-standing Poincaré Conjecture in five or more dimensions. Also, three decades ago, he played a key role in the early mathematical development of the theory of chaotic dynamical systems. At that time he was investigating the long-run behavior of dynamical systems: What does the system settle down to as time goes on indefinitely? Originally he hoped to prove that for a "typical" system the only possibilities are that it settles down to a steady state or a stable limit cycle. But after being shown a counterexample, he discovered what can give many typical systems a strange, chaotic attractor: the Smale horseshoe. In his lecture he describes for us the workings of the horseshoe, and he identifies the crux of chaos: homoclinic points (explained in his lecture).

John Polkinghorne has had two careers: the first as a physicist, the second as an ordained theologian in the Church of England. By tradition each Nobel Conference includes on its panel at least one speaker whose assignment is to address the philosophical or theological implications of the conference theme. We feel particularly fortunate that we were able to invite a scholar who has solid credentials in both science and theology, a person who has a thorough understanding of the rapidly developing science of chaos and a deep grounding in metaphysics and theology. Polkinghorne regards fundamental theology as "the great integrating discipline," and in his lecture, "Chaos and Cosmos: A Theological Approach," he outlines a theology illuminated by the insights of chaos theory. His approach provides a new perspective on some great issues of metaphysics and theology: determinism and free will, the mind-body problem, and the relationship of God to the laws of physics.

James Gleick, author of the best-selling book *Chaos: Making a New Science*, has been instrumental in informing the public of the potentially revolutionary contributions of chaos theory to our scientific world view. A writer rather than a scientist, he is a master of metaphor who applies his talent to the lucid exposition of difficult new scientific ideas. As the final speaker at the conference, he invites us to re-examine our metaphors for natural processes, enlightened by the discoveries in chaos and fractals, and he offers his own reflections on the conference proceedings and his projections for the future of chaos. In the light of the new science of chaos he challenges us to reconsider what the fundamental laws of nature really are. Are they the laws of the fundamental particles, from which, in principle but not in practice, all other phenomena may be explained? Or are they laws about complex systems, relating order and disorder, organization and dissolution, life and death?

It is in the hope that the new science of chaos can help us all to understand and answer the questions about nature and life that are the most interesting, that really matter to us, that we recommend this book to our readers.

St. Peter, Minnesota
August 1992

CONTRIBUTORS

Benoit Mandelbrot
IBM fellow, IBM Thomas J. Watson Research Center, Yorktown Heights, New York, (1974-); IBM staff research member (1958-1974); Abraham Robinson Adjunct Professor of Mathematical Sciences, Yale University (1987-); fellow or member of numerous scientific societies; editorial board, **Advances in Applied Mathematics** (1984-) and **Journal of Visual Communication and Image Representation** (1989-); author or editor of articles, conference proceedings and books, including **Fractals: Form, Chance and Dimension** (1977), **The Fractal Geometry of Nature** (1982) and **Fractals and Multifractals: Noise, Turbulence, and Galaxies** (in preparation).

Heinz-Otto Peitgen
Professor of mathematics, University of Bremen (1977-); professor of mathematics, University of California, Santa Cruz (1985-1991) and Florida Atlantic University (1991-); cofounder, Institute for Dynamical Systems at Bremen; editorial board, **Acta Applicandae Mathematicae, Dynamics Reported**, and **Bifurcation and Chaos**; author of articles, catalogs and books, including **The Beauty of Fractals** (1986), **The Science of Fractal Images** (1988), **Fractals for the Classroom** (1991); co-creator of several scientific movies and exhibitions of computer-generated art currently touring internationally.

Mitchell Feigenbaum
Professor of mathematics and physics, Rockefeller University (1986-); laboratory fellow, Los Alamos National Laboratory (1981); staff member (theoretical physics), Los Alamos National Laboratory (1974-1982); editorial board, **Journal of Statistical Physics** (1981); member, Sigma Xi, New York Academy of Sciences; author of articles published in various professional journals, including **Journal of Statistical Physics, Nonlinearity,** and **American Journal of Physics.**

Ilya Prigogine
Director of the International Institute of Physics and Chemistry, Solway (1959-); director of the Ilya Prigogine Center for Studies in Statistical Mechanics, Thermodynamics and Complex Systems (University of Texas) (1967-); Ashbel Smith Regents Professor, University of Texas (1984-); Nobel Prize in Chemistry (1977); member, honorary member or fellow of numerous learned societies, including the World Academy of Art and Sciences, the Belgian Royal Academy of Sciences, Letters and Fine Arts and the International Academy of the Philosophy of Science; author of works including **Thermodynamic Theory of Structure, Stability and Fluctuations** (1971), **From Being to Becoming: Time and Complexity in the Physical Sciences** (1980) and **Order Out of Chaos** (1983).

Steve Smale
Professor of mathematics, University of California, Berkeley (1964-); member, American Academy of Arts and Sciences, National Academy of Sciences; fellow, the Econometric Society; awarded the Veblen Prize for Geometry of the American Mathematical Society (1965), Fields Medal, International Union of Mathematics (1966) and the Chauvenet Prize of the Mathematical Association of America (1988); author of numerous articles and papers.

John Polkinghorne
President, Queen's College, Cambridge (1989-); Fellow, Dean and Chaplain of Trinity Hall, Cambridge (1986-1989); Lecturer (1958-1965), Reader (1965-1968), Professor of Mathematical Physics (1968-1979), University of Cambridge; Fellow of the Royal Society; member, Church of England Doctrine Commission (1989-); author of many papers on theoretical elementary particle physics and books, including **The Particle Play** (1979), **The Way the World Is** (1983), **Science and Creation** (1988), and **Science and Providence** (1989).

James Gleick
McGraw Distinguished Lecturer, Princeton University (1989-); contributor, **New York Times Magazine;** editor, metropolitan news and science reporter, **The New York Times** (1978-1988); author of **Chaos: Making a**

New Science (1987 National Book Award and Pulitzer Prize nominee in general nonfiction).

John Holte
Professor of mathematics and computer science, Gustavus Adolphus College (1976-); M.S., Ph.D. the University of Wisconsin; B.S., California Institute of Technology; author of expository and research papers on probability theory and fractals; chair of 1990 Nobel Conference.

FRACTALS
B.B. Mandelbrot

Fractals are a family of geometric shapes, and I happen to believe that, in order to understand geometric shapes, one must see them. It has very often been forgotten that geometry simply *must* have a visual component, and I believe that in many contexts this omission proved to be very harmful. Therefore, I shall show you a large number of slides.*

But let me begin with a kind of abstract without slides by saying a few general words concerning the scope of fractal geometry. I view it as a workable geometric middle ground between the excessive geometric order of Euclid and the geometric chaos of general mathematics. It is based on a form of symmetry that had previously been underutilized, namely self-similarity, or some more general invariance under contraction or dilation.

Fractal geometry is conveniently viewed as a language, and it has proven its value by its uses. Its uses in art and pure mathematics, being without "practical" application, can be said to be poetic. Its uses in various areas of the study of materials and of other areas of engineering are examples of practical prose. Its uses in physical theory, especially in conjunction with the basic equations of mathematical physics, combine poetry and high prose. Several of the problems that fractal geometry tackles involve old mysteries, some of them already known to primitive man, others mentioned in the Bible, and others familiar to every landscape artist.

This being granted, let me elaborate. Let me first remind you of a marvelous text that Galileo wrote at the dawn of science:

"Philosophy is written in this great book — I am speaking of the Universe — which is constantly offered for our contemplation, but which

---

* It is unfortunately impossible to reproduce color slides in this written version of my talk; hence the text had to be recast to fit a far smaller number of black and white figures. It is also impossible to insert here a video tape that was part of my talk. It is based on a *Fractals and Music* event, which I prepared with the composer Charles Wuorinen. It was first presented in April 1990 at the Guggenheim Museum in New York, and then in 1991 at the Alice Tully Hall of Lincoln Center in New York. This tape was a convenient way of presenting a large number of fractals in a short time.

cannot be read until we have learned its language and have become familiar with the characters in which it is written. It is written in the language of mathematics, and its characters are triangles, circles and other geometric forms, without which it is humanly impossible to understand a single word of it; without which one wanders in vain across a dark labyrinth." (Galileo Galilei: *Il Saggiatore*, 1623).

We all know that mechanics and calculus and, therefore, all of quantitative science, were built on these characters, and we all know that these characters belong to Euclidean geometry. In addition, we all agree with Galileo that this geometry is necessary to describe the world around us, beginning with the motion of planets and the fall of stones on Earth.

But is it sufficient? To answer, let us focus on that part of the world that we see in everyday life. Modern boxlike buildings are cubes or parallelepipeds. Good plasterboard is flat. Good quality tables are flat and typically have straight or circular edges. More generally, it is typical for the works of Man the Engineer and Builder to be flat, round, or to follow other very simple shapes of classical school geometry.

By contrast, many shapes of nature — for example, the shapes of mountains, clouds, broken stones, and trees — are far too complicated for Euclidean geometry. Mountains are not cones. Clouds are not spheres. Island coastlines are not circles. Rivers don't flow straight. Therefore, we must go beyond Euclid if we want to extend science to those aspects of nature.

A geometry able to include mountains and clouds does exist now. I put it together in 1975, but of course it incorporates very many pieces that have been around for a very long time. Like everything in science, this new geometry has very, very deep and long roots. Let me illustrate some of the tasks it can perform.

Figure 1. A fractal landscape that never was (R.F. Voss)

Figure 1 seems to represent a real mountain, but it is neither a photograph nor a painting. It is a mathematical forgery, a computer forgery; it is completely based upon a mathematical formula from fractal geometry.

The same is true of the forgery of a cloud that is shown in Figure 2.

An amusing and important feature of Figures 1 and 2 is that both are adaptations of formulas that had been known in pure mathematics. Thanks to fractal geometry, diverse mathematical objects which used to be viewed as being very far from physics have turned out to be the proper tools for studying nature. I shall return to this in a moment.

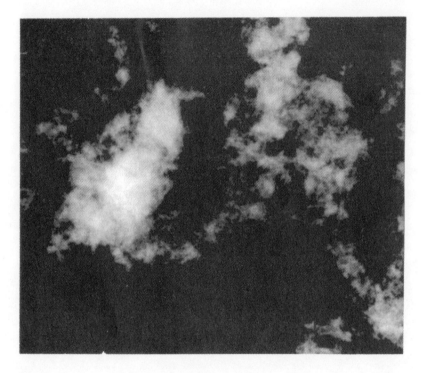

Figure 2. A cloud formation that never was (S. Lovejoy & B.B. Mandelbrot)

Fractal modeling of relief was successful in an unexpected way. It is used in an immortal masterpiece of cinematography called *Star Trek II: The Wrath of Khan*. Many people have seen it, but very few have noticed without prodding that the new planet that appears in the *Genesis* sequences of that film is fractal. If I could show it to you, you would see that it happens to have peculiar characteristics (superhighways and square fields). They are due to a shortcut taken by Lucasfilm to make it possible to compute these fractals quickly enough. But we need not dwell on flaws. Far more interesting is the fact that the films that include fractals create a bridge between two activities that are not expected to ever meet, mathematics and physics on the one hand, and popular art on the other.

More generally, an aspect of fractals that I found very surprising at the beginning, and that continues to be a source of marvel, is that people respond to fractals in a deep emotional fashion. They either like them or dislike them, but in either case that emotion is completely at variance with the boredom that most people feel toward classical geometry. Let me state that I will never say anything bad about Euclid's geometry. I love it. It was an important part of my life as a child and as a student; in fact, the main reason why I survived academically despite chaotic schooling was I could always use geometric intuition to cover my lack of skill as a manipulator of formulas. But we all know by experience that almost everybody except professional geometers views Euclid as being cold and dry. The fractal shapes are exactly as geometric as those of Euclid, yet they evoke emotions that geometry is not expected or supposed to evoke.

Now a few preliminary words about deterministic chaos. This is a topic that I and other speakers will describe later, but something should be mentioned immediately. The proper geometry of deterministic chaos is the same as the proper geometry of mountains and clouds. The fact that we need only one new geometry is really quite marvelous, because several might have been needed, in addition to that of Euclid. But it is not so. Fractal geometry plays both roles. Not only is it the proper language to describe the shape of mountains and of clouds, but in addition it is the proper language for all the geometric aspects of chaos, which you will hear discussed by several of speakers who will follow.

I have myself devoted much effort to the study of deterministic chaos, and would like to show you a few examples of the shapes I have encountered in this context.

Figure 3 is an enormously magnified fragment from a set to which my name has been attached. Here a fragment of the Mandelbrot set has been magnified in a ratio equal to Avogadro's number, which is about 10^{23}. Why choose this particular number? Because it is a nice very large number, and a huge magnification provided a good opportunity for testing the quadruple precision arithmetic on the I.B.M. computers of a few years ago. (They passed the test. It is very amusing to be able to justify plain fun and pure science on the basis of such down-to-earth specific roles.) If the whole

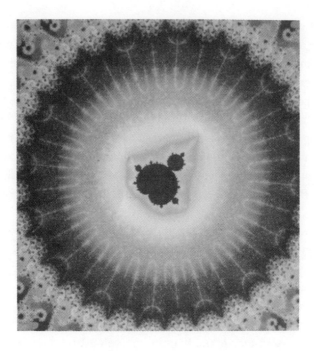

Figure 3. A very small fragment of the Mandelbrot set (R.F. Voss)

Mandelbrot set had been drawn on the same scale, the end of it would be somewhere near the star Sirius.

The shape of the black "bug" near the center is very nearly the same as that of the white center of Figure 12, which will show the shape of the whole Mandelbrot set. Finding bugs all over is a token of geometric orderliness. On the other hand, the surrounding pattern is not present in the whole set, but depends very much upon the point on which the zoom has focused (this is a token of variety, and even chaos). I shall return to this set later.

The shape shown in Figure 4 is a variant of the Mandelbrot set that corresponds to a slightly different formula. This shape is reproduced here simply to comment on a totally amazing and extraordinarily satisfying aspect of fractal geometry. Fractals are perceived by many people as being beautiful. But these shapes were initially developed for the purpose of

Figure 4. A small fragment of a modified Mandelbrot set (B.B. Mandelbrot)

science, for the purpose of understanding how the world is put together both statically (in terms of mountains) and dynamically (in terms of chaos, strange attractors, etc.). In other words, the shapes shown in Figures 1 to 4 were not *intended* to be beautiful. This being beautiful unavoidably raises many questions. The most important question is simply, why are they beautiful? This fact must tell us something about our system of visual perception.

I wanted to start with Figures 1 to 4 because their structure is rich. But I went overboard. Their structure is, in fact, so rich that these figures cannot be used to explain the main feature of all fractals. The underlying basic

Figure 5. Cauliflower *Romanesco* (R. Ishikawa)

principle shows far more clearly in Figure 5, which—for a change—reproduces a real photograph of a real object. You may recognize a variety of cauliflower called *Romanesco*. Each bud looks absolutely like the whole head, and in turn each bud sub-divides into smaller buds, and so on. I am told that the same structure repeats, over five levels of separation you can do by hand and see by the naked eye, and then over many more levels you can only see with a magnifying glass or microscope.

Till recently, scientists did not pay much attention to this "hierarchical" property. Their first reaction to this kind of botanical shape was not to focus on buds within buds, but on the spirals formed by the buds. This interest led to extensive knowledge about the relation between the golden mean (and the

Fibonacci series) and the way plants spiral. But the hierarchical structure of buds is more important for us here, because it embodies the essential idea of fractal.

Before we continue and tackle what a fractal *is*, let us ponder what a fractal *is not*. Take a geometric shape and examine it in increasing detail. That is, take smaller and smaller portions near a point P, and allow every one to be dilated, that is, enlarged to some prescribed overall size.

If our shape belongs to standard geometry, it is well known that the enlargements become increasingly smooth. For example, one can say that one expects a curve to be "attracted" under dilations to a straight line (thus defining the tangent at the point P). The term "attractor" is borrowed from dynamics and probability theory. One also expects a curve to be attracted under dilation to a plane (thus defining the tangent plane at the point P).

More generally, one can say that nearly every standard shape's local structure converges under dilation to one of the small number of "universal attractors." The grandiose term, "universal," is borrowed from recent physics. An example of an exception to this rule is when P is a double point of a curve; the curve near P is then attracted to two intersecting lines and has two tangents; but double points are few and far between in standard curves.

Yet the shapes I have been showing *fail to be* locally linear. In fact, they deserve being called "geometrically chaotic," unless proven otherwise. In an altogether different neighborhood, in the great City of Science, a kind of geometric chaos became known during the half century from 1875 to 1925. Then, while attempting to flee from concern with nature, mathematicians became aware of the fact that a geometric shape's roughness *need not* vanish as the examination becomes more searching. It is conceivable that it could either remain constant, or vary endlessly, up and down. The hold of standard geometry was so powerful, however, that the resulting shapes were not recognized as models of nature. Quite to the contrary, they were labeled "monstrous" and "pathological." After discovering these sets, mathematics proceeded to increasingly great generality.

Science must constantly navigate between two dangers: lack and excess of generality. Between the two extremes it must always find the proper level that is necessary in order to do things right. Between the proper level that is necessary in order to do things right. Between the extremes of the excessive geometric order of Euclid and of the geometric chaos of the most general mathematics, can there be a middle ground of "organized" or "orderly" geometric chaos? To provide such a middle ground is the ambition of fractal geometry.

The reason why fractals are far more special than the most general shapes of mathematics is that they are characterized by so-called "symmetries," which are invariances under dilations and/or contractions. Broadly speaking, mathematical and natural fractals are shapes whose roughness and fragmentation *neither* tend to vanish *nor* fluctuate up and down, but remain *essentially unchanged* as one zooms in continually and examination is refined. Hence, the structure of every piece holds the key to the whole structure.

The preceding statement is made precise and illustrated by Figure 6, which represents a shape that is enormously simpler than Figures 1 to 5. As a joke, I called it the *Sierpiński gasket,* and the joke has stuck.

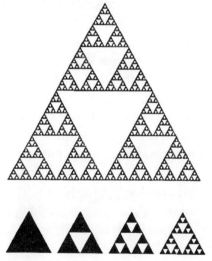

Figure 6. The Sierpiński gasket. Early and late states of construction

The four small diagrams show the point of departure of the construction, then its first three stages, while the large diagram shows an advanced stage. The basic step of the construction is to divide a given (black) triangle into four subtriangles, and then erase (whiten) the middle fourth. This step is first performed with a wholly black filled-in triangle of side 1, then with three remaining black triangles of side 1/2. This process continues, following a pattern called *recursive deletion*, which is very widely used to construct fractals. Related patterns are *recursive substitution* and *recursive addition* (which we shall encounter soon) and *recursive multiplication* (which is beyond the scope of this talk).

Now, take the gasket and perform an isotropic linear reduction whose ratio is the same in all directions—namely, *1/2*—and whose fixed point is any one of the three apexes of the triangle that circumscribes the gasket. This transformation is called a *similarity* (or, more precisely, a *homothety*). By examining the large advanced-stage picture, it is obvious that each of the three reduced gaskets is simply superposed on one-third of the overall shape. For this reason, the fractal gasket is said to have the properties of *self-similarity*—more precisely, of *exact*, or *linear*, self-similarity.

Those heavy terms are needed today because one can also understand "similar" as a loose everyday synonym of "analogous." In the early days of fractal geometry, the resulting terminological ambiguity was acceptable to physicists, because early detailed studies did indeed concentrate on strictly self-similar shapes. However, more recent developments have extended to include *self-affine* shapes, in which the reductions are still linear, but the reduction ratios in different directions are different. For example, a relief is nearly self-affine: in order to go from a large piece to a small piece, one must contract the horizontal and vertical coordinates in different ratios.

When the Sierpiński gasket is constructed by deleting middle triangles, as in Figure 6, its properties of contracting self-similarity appear, so to speak, as "static" and "after-the-fact." But this is a completely misleading impression. Its prevalence and its being viewed as a flaw continually surprise me. In fact, the same symmetries can be reinterpreted "dynamically," and they suffice to generate the gasket. The device, which is called the "chaos game," is a stochastic interpretation of a scheme due to Hutchinson.

Start with an "initiator" that is an arbitrary bounded set, for example, a point P_0. Denote the three similarities of the gasket by S_0, S_1, and S_2, and denote by $k(m)$ a random sequence of the digits 0, 1, and 2. Then define an "orbit" as made of the points $P_1 = S_{k(1)}(P_0)$, $P_2 = S_{k(2)}(P_1)$, and more generally $P_j = S_{k(j)}(P_{j-1})$. One finds that this orbit is "attracted" to the gasket, and that after a few stages it describes its shape very well.

I used to think that the word "self-similarity" was used for the first time in a paper of mine in 1964. But it has since come to my attention that at least one reference had used it once before me. Now, why was this word never used, though the idea itself is perfectly obvious and must be very old? The reason is that the shapes to which it refers had no importance until my work. For example, Sierpiński had defined his shape for some purpose which has long been forgotten—because it was not very important.

Why did self-similarity become important? One reason why fractal geometry has taken an enormous purview, and I spent so much time in efforts to build it as a discipline, resides in empirical discoveries (each established by a separate investigation) that the relief of the earth is self-similar, and that the same is true of many other shapes around us. Figures 1 to 5 suffice to show that the impression that self-similarity is a barren and not very fruitful idea would be an altogether wrong impression.

Granted what has just been said, why did the gasket become important? It does not represent anything of interest; in fact, it is so relentlessly monotonous, that it could be called as simple as Euclid. In a few days of study, you will know nearly everything about it. The same holds for another widely known shape, called the snowflake curve, or Von Koch Island, for a set called Cantor dust, and for a few other long-known structures of the same ilk. The reason why they are important is that you must begin the study of fractal geometry *with* the Sierpiński gasket and its ilk. But keep in mind that the real fun begins *beyond* them.

The fun begins after one has added an element of unpredictability, which may be due to either randomness (as in Figures 1, 2, and 5), or to non-linearity (as in Figures 3 and 4). Non-linearity is the key word of the new meaning of chaos, namely, of deterministic chaos, and randomness is the key to chaos in the old sense of the word. The two are very intimately linked.

Figure 7. Mandelbrot's Peano curve (B.B. Mandelbrot)

But let us not rush away from linearly self-similar fractals, because in some cases a suitable graphic rendering suffices to break their relentless monotony.

Figure 7 shows my variant of a curve that Giuseppe Peano constructed in 1890. The point of Peano curves is that they manage to fill a portion of the plane. Pages and pages have been written by mathematicians to praise the freedom of imagination that allows Man to invent shapes that are so absolutely removed from reality. The Peano curve was specifically designed to be a counterexample to a natural belief that used to be universal: that curves and surfaces do not mix. It was designed for the purpose of separating mathematics and physics into two completely independent investigations. Unfortunately, it was quite successful in that respect, at least for a century.

To obtain my new Peano curve, you replace an initial straight segment by the complicated zigzag (top left). Then (top middle) each zig and zag is replaced by a smaller version of the zigzag on the top left. The same pattern (called *recursive substitution*) is then repeated without end. The top right diagram makes it easy to believe that the boundary between black and white will end up by filling a snowflake curve. I call it "snowflake sweep." The bottom of Figure 7 is just a fancy computer rendering of my curve, in which every segment is replaced by an arc of circle. Everybody can see in it all kinds of branching systems of arteries and veins, or of rivers, or of flames, or whatever. It is very difficult not to see very realistic things in it. But they were not seen until my work, and then (partly as a result) mathematics and physics did indeed move in very different directions.

To continue to demolish the impression of the "inadequacy" of self-similarity, Figure 8 combines a sequence of completely artificial random landscapes. Each part of this picture consists in enlarging a small black rectangle in the preceding picture, and in filling in additional detail. This procedure is called *recursive addition*. Each step yields a landscape that is of course different from the preceding landscape. It is more detailed, yet at the same time is qualitatively the same. The successive enlargements might have been different parts of the same coastline examined on the same scale, but in fact they are neighborhoods of one single point examined at very different scales. Clearly, these successive enlargements of a coastline completely fail to converge to a limit tangent!

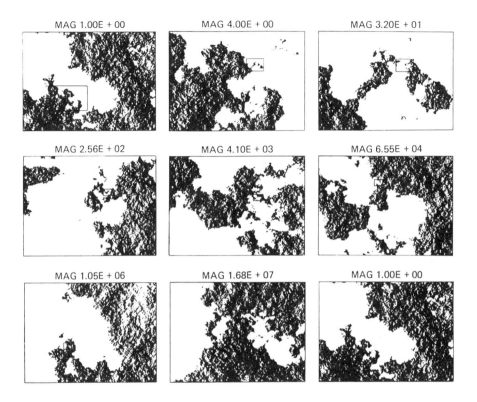

Figure 8. Zoom onto a fractal landscape that never was (R.F. Voss)

At this point, let me recall a story about the great difficulties the ancient Greeks used to experience in defining "size." Navigators knew that Sardinia took longer to circumnavigate than Sicily. On the other hand, there was evidence that Sardinia's fields are smaller than Sicily's.

So which was the bigger island? The Greeks, being sailors above all, seem to have long held the belief that Sardinia was the bigger of the two, because its coastline was longer. But let us examine Figure 9 and ponder the notion of the length of a coastline. When the ship used to circumnavigate is large, the captain will report a rather small length. A much smaller ship would come closer to the shore, and navigate along a longer curve. A man walking along the coastline will measure an even longer length. So what about the "real length of the coast of Sardinia"? The question seems both elementary and silly, but it turns out to have an unexpected answer. The answer is, "It depends." The length of a coastline depends on whether you circumnavigate in a large or a small ship, or walk along it, or use a mouse or some other instrument to measure the coastline.

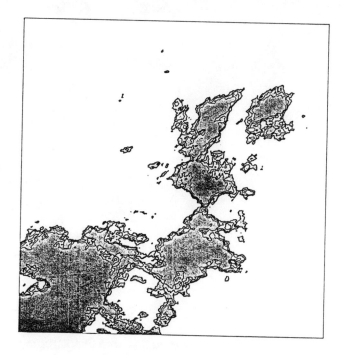

Figure 9. A fractal coastline that never was (B.B. Mandelbrot)

This makes us appreciate the extraordinary power of the mental structure that schools have imposed by teaching Euclid. Many people who thought they had never understood geometry have nevertheless learned enough to expect every curve to have a length. For the curves in which I am interested, this turns out to have been the wrong thing to remember from school, because the theoretical length is infinite, and the practical length depends on the method of measurement. Its increase is faster where the coastline is rough, making it necessary to study the notion of roughness.

This last notion is fundamental, because the world we live in includes many rough objects, and therefore Man must learn to avoid them to avoid great harm. However, the task of measuring roughness objectively has turned out to be extraordinarily difficult. People whose work demands it, like metallurgists, seem to ask their friends in statistics for a number one could measure and call roughness. But perform the following experiment. Take different samples of steel that the US National Bureau of Standards guarantees to be pieces of one block of metal, as homogenous as man can make it. If you take several pieces and you break them all and measure the roughness of the fractures according to the books on statistics, you will get values which are in complete in disagreement.

On the other hand, I shall argue that roughness happens to be measured consistently by a quantity called *fractal dimension*, which happens in general to be a fraction, and which one can measure very accurately. Studying many samples from the same block of metal, we found the same dimension for every sample.

To understand the idea that fractal dimension is a proper measure of roughness, let us recall what is meant by saying that temperature is a proper measure of the notion of hotness. Man must have known forever that some things are hot and other things are cold, but physics required a reliable way of measuring hotness. This came only when the thermometer was invented, and different people using the same thermometer could get the same value of hotness for the same object. A description of the degree of hotness by one number was a necessary first step before physics could move on to a theory of matter.

Similarly and most fortunately, fractal geometry started with a few ideas about how to express roughness and complexity by a number. Some of these ideas add up a bunch of related but distinct tools (one can think of them as being different types of screwdrivers) that are collectively called "fractal dimensions." People who work with fractal geometry quickly develop an intuition of fractal dimension and can now guess it very accurately for the simple shapes.

The reason for the term "dimension" is that these notions' definitions can also be applied to points, intervals, full squares, and full cubes, and in those cases yield the familiar values 0, 1, 2, and 3. Applied to fractals, however, these definitions usually yield values that are not integers. The loose idea of "roughness" has turned out to demand a number of distinct numerical implementations; hence the multiplicity of distinct "fractal dimensions" has proven valuable. A dimension due to Hausdorff and Besicovitch was the first example, but for practical needs it is either too difficult or too specialized.

The simplest variant is the *similarity* dimension D_S, which applies to shapes that are linearly self-similar. As I have already said, this means that they are made up of N replicas of the whole, each replica being reduced linearly in the same ratio r. Then one defines

$$D_S = \frac{\log N}{\log (1/r)}.$$

For a point, an interval, a square, and a full cube, one has $D_S = 0, 1, 2,$ and 3. As announced, these are the familiar values of the "ordinary" dimensions. But for the Sierpiński gasket we have $N = 3$ and $r = 1/2$, hence $D = \log 3/\log 2 \sim 1.5849$.

Another simple fractal dimension is the *mass dimension*. Take a distribution of mass of uniform density on the line, in the plane, or in space. Then choose a sphere of radius R whose center lies in our set. The mass in such a sphere takes the form $M(R) = FR^D$, where D is the "ordinary" dimension and F is a numerical constant. The idea of uniform density extends to fractals, and in many cases an exponent D can be defined; it is called the *mass dimension* and is often equal to the similarity dimension.

This must, unfortunately, be enough for dimension. The next topic I wish to tackle is the increasingly valuable role of fractal geometry as a tool in the discovery and the study of new aspects of nature. To illustrate, I have chosen to sketch for you a form of random growth that generates the *fractal diffusion limited aggregates,* or Witten-Sander aggregates. A DLA cluster lurks in the center of Figure 10. It is a tree-like shape of baffling complexity one can use to model how ash forms, how water seeps through rock, how cracks spread in a solid, and how lightning discharges.

Figure 10. A cluster of diffusion limited aggregation, surrounded by its equipotential curves (C.J.G. Evertsz and B. B. Mandelbrot)

To see how the growth proceeds, take a very large chess board and put a queen, which is not allowed to move, in the central square. Pawns, which are allowed to move in any of the four directions on the board, are released from a random starting point at the edge of the board, and are instructed to perform a random walk, or drunkard's walk. The direction of each step is chosen from four equal probabilities. When a pawn reaches a square next to that of the original queen, it transforms itself into a new queen and cannot move any farther. Eventually one has a branched, rather spidery-looking, collection of queens.

Quite unexpectedly, massive computer simulations have shown that DLA clusters are fractal. They are nearly self-similar, that is, small portions are very much like reduced versions of large portions. But clusters deviate from randomized linear self-similarity, something that will pose interesting challenges for the future.

One reason for the importance of DLA is that it concerns the interface between the smooth and the fractal. A premise of fractal geometry is that much in the world is fractal. But science is expected to be cumulative, the new being added to the old without chasing it away. Therefore, the new wisdom must not deny the old scientific wisdom, whose premise it is that the world is made of smooth shapes and involves smooth variation and differential equations.

Let us describe how DLA shows that the old and the new wisdom are compatible if one abandons the old philosophical expectation that *everything* in the world will eventually prove to be smooth or of smooth variation.

To show how smooth variation can produce rugged behavior, the original construction must first be rephrased in terms of the theory of electrostatic potential. The description that follows is necessarily a little schematic. Take the big box in which DLA is growing, and connect it to a positive potential, to be taken as unity, and connect the cluster to the potential *0*. Then the value of the potential elsewhere in the box is best described by equipotential curves, for example, the curves along which the potential takes the increasing values. *01, .02, ..., .99*. Figure 10 shows that all these curves are smooth, and that they provide a progressive transition between the box and the boundary of the cluster. Analytic calculation is out of the question, but "physical common sense" can be combined with numerical calculation. In effect, the object's boundary includes many needles, and each has a high probability of getting hit by lightning. This is manifested by equipotential lines becoming crowded together near the tips of a DLA cluster. More generally, returning to the random pawns that build up a DLA cluster, the position where the pawn lands is obtained from the shapes of the electrostatic equipotentials.

Now we come to the next logical step, which implies that DLA has brought an intellectual innovation of the highest order. For nearly 200 years,

Now we come to the next logical step, which implies that DLA has brought an intellectual innovation of the highest order. For nearly 200 years, the study of potentials has involved fixed boundaries. But in the simple random walk that creates DLA, a "hit" in the above terminology can be interpreted as provoking a displacement of the boundary. Thus, the massive numerical experiments about DLA teach us that, when one allows boundaries to move in response to the potential, the boundaries become fractal.

This implies that we know without any trace of doubt that one can create rough fractals from the smoothness characteristic of equipotential lines. But this knowledge remains imperfect. We would clearly like to add some exact mathematics and some additional physical argument. Nevertheless, it is worth noting how fractal geometry has led to an altogether new problem, has outlined the broad line of solution, and has set many scientists to work.

Now I am going to move for a second time away from randomness to deterministic chaos, and from objects in real physical space to imaginary objects. What will remain unchanged is that we shall deal with spiky sets surrounded by smooth equipotential lines.

The first notion here is that of a Julia set of quadratic iteration. Pick a point C with coordinates u and v, and call it a "parameter." Next pick, in a different plane, a point P_0 with coordinates x_0 and y_0. Then form $x_1 = x_0^2 - y_0^2 + u$ and $y_1 = 2x_0y_0 + v$. These formulas may seem a bit artificial, but they simplify if the point C with coordinates x and y is represented by a complex number $z = x + iy$. (One can add and multiply complex numbers like ordinary numbers, except that i^2 must always be replaced by -1.) In terms of the complex numbers $C = u + iv$ and $z = x + iy$, the preceding rule simplifies to $z_1 = z_0^2 + C$ and (more generally) $z_{k+1} = z_k^2 + C$. But even the reader who is scared of complex numbers will understand the expressions in terms of x_k and y_k.

When the orbit P_k fails to escape to infinity, the initial P_0 is said to belong to the "filled-in Julia set." An example is shown in Figure 11. If you start outside of the black shape, you go to infinity. If you start inside, you fail to iterate to infinity.

The boundary between black and white is called Julia curve. It is approximately self-similar. Each chunk is not quite identical to a bigger

Figure 11. Quadratic Julia sets for the map $z \rightarrow z^2 + C$. Each boundary of a zebra strip corresponds to a different value of C (B.B. Mandelbrot)

chunk, because of non-linear deformation. But it is astonishing that iteration should create any form of self-similarity, quite spontaneously.

As in the investigation of fractal mountains, the computer was essential to the study of iteration. The bulk of fractal geometry is concerned with shapes of great apparent complication, and by hand they could never be drawn. More precisely, this picture might have been computed by a hundred different people working for years. But nobody would have started such an enormous calculation without first feeling that it was worth performing.

Not only had I access to a computer in 1979, but I was familiar with its power. Therefore I felt these calculations were worth trying, even though I certainly did not know what was going to come out. A fishing expedition led to a primitive form of Figure 12. The Julia sets of the map $z^2 + C$ can take all kinds of shapes, and a small change in C can change the Julia set very greatly. I set out to classify all the possible shapes (for reasons which I have no time to discuss) and came up with a new shape. It is called the *Mandelbrot set*, which of course I find to be a great honor. Figure 3, which appeared earlier, was a tiny portion of Figure 12.

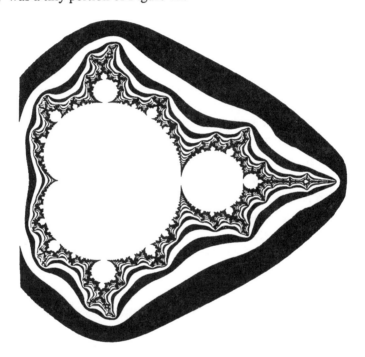

Figure 12. The Mandelbrot set, surrounded by its equipotential curves

Here is how the M set is constructed. Take a starting point C_0 in the plane with coordinates u_0 and v_0. From the coordinates of C_0, form a second point C_1 with coordinates $u_1 = u_0^2 - v_0^2 + u_0$ and $v_1 = 2u_0u_0 + v_0$. Next, form the point C_2 of coordinates $u_2 = u_1^2 - v_1^2 + u_0$ and $v_2 = 2u_1v_1 + v_0$. More generally, the coordinates u_k and v_k of C_k are obtained from u_{k-1} and

v_{k-1} by the so-called "iterative formulas" $u_k = u_{k-1}^2 - v_{k-1}^2 + u_0$ and $v_k = 2u_{k-1}v_{k-1} + v_0$. When C_0 is represented by $z_0 = u_0 + iv_0$, the above formulas simplify to $z_1 = z_0^2 + z_0$, and $z_k = z_{k-1}^2 + z_0$. The points C_k are said to form the orbit of C_0, and the set M is defined as follows: If the orbit C_k fails to go to infinity, one says that C_0 is contained within the set M. If the orbit C_k does go to infinity, one says that the point C_0 is outside M.

This algorithm concerns the following very sober problem of deterministic dynamics. When C_0 is in the interior of M, quadratic dynamics yields an orbit that is perfectly orderly, in the sense that it is asymptotically periodic. When C_0 is outside M, to the contrary, the behavior of the orbit is deterministic but practically unpredictable, hence is chaotic. Quadratic dynamics was singled out for detailed study because in this case the criterion separating orderly from chaotic behavior is, as seen above, as clean as can be. The boundary between the two possibilities turns out to be messy beyond any expectation.

As you zoom towards a portion of the boundary of M, part of what you see is simply a repetition of something you have already seen. This element of repetition is essential to beauty. But beauty also requires an element of change, and you get it also very clearly. As you come closer and closer, what you see becomes more and more complicated. The overall shape is the same, but the hair structure becomes more and more intense. This feature is *not* something we put in on purpose. Insofar as mathematics is not invented but discovered, it is something that has been there forever, and it shows that the mathematics of z squared plus C is astonishingly complicated by contrast with the simplicity of the formula. We find that the M set, when examined closer and closer and closer, exhibits the co-existence of the relentless repetition of the same theme combined with variety that boggles the imagination. I first saw the Mandelbrot set on a black and white screen of very low graphic quality, and the picture looked dirty. But when we zoomed on what seemed like dirt we found instead an extraordinary little copy of the whole.

In Figure 12, the Mandelbrot set is the *white* "bug" in the middle. It is very rough-edged, but is surrounded by a collection of zebra stripes whose edges become increasingly smooth as one goes away from M. These zebra stripe edges happen to be Laplacian equipotential curves — just as in Figure 10. But they are far easier to obtain.

Of course, the black-and-white figures in this paper are far from the beautiful color ones which everyone must have seen. The quality of the color rendering of the Mandelbrot set shows the skills of the programmers, as does as the quality of the color rendering of fractal mountains, but the M-set structure itself is independent of the color rendering. What is important is that the structure is so complicated that we could not understand it unless the color rendering were sufficiently rich. In fact, the set has such an enormous amount of structure that we cannot see in one single color rendering. Different renderings emphasize very different aspects of it. Again, this structure was *not* invented for the purpose of doing something beautiful, but purely for the purpose of exploring the advanced theory of z squared plus C.

To the layman, fractal art tends to seem simply magical, but no mathematician can fail to try to understand its structure and its meaning. A remarkable aspect of recent events is that much of the mathematics triggered by the M set, had its visual origin been hidden, could have passed as "pure." To many mathematicians, the newly opened possibility of playing with pictures interactively has turned out to reveal a new mine of purely mathematical questions and of conjectures, of isolated problems, and of whole theories. To take an example, examination of the Mandelbrot set led me in 1980 to many conjectures which were simple to state, but then proved very hard to crack. To mathematicians, their being difficult and slow to develop does not make them any less fascinating, because a host of intrinsically interesting "side-results" have been obtained in their study.

Herein hangs a tale. Pure mathematics certainly does exist as one of the remarkable activities of Man, it certainly is different in spirit from the art of creating pictures by numerical manipulation, and it has indeed proven that it can thrive in splendid isolation—at least over some brief periods. Nevertheless, the interaction between art, mathematics, and fractals confirms what is suggested by almost all earlier experiences. Over the long haul mathematics gains by not attempting to destroy the "organic" unity that appears to exist between seemingly disparate but equally worthy activities of Man, the abstract and the intuitive.

Let me now bring together the separate strings of my talk. How did fractals come to play their role of "extracting order out of chaos"? The key resides in the following very surprising discovery I made thanks to computer graphics.

The algorithms that generate fractals are typically so extraordinarily short as to look positively dumb. This means they must be called "simple." Their fractal outputs, to the contrary, often appear to involve structures of great richness. A priori one would have expected that the construction of complex shapes would necessitate complex rules.

What is the special feature that makes fractal geometry perform in such unusual manner? The answer is very simple. The algorithms are recursive, and the computer code written to represent them involves "loops." That is, the basic instructions are simple, and their effects can be followed easily. But let these simple instructions be followed repeatedly. Unless one deals with the simple old fractals (Cantor set or Sierpinski gasket), the process of iteration effectively builds up an increasingly complicated transform, whose effects the mind can follow less and less easily. Eventually one reaches something that is "qualitatively" different from the original building block. One can say that the situation is a fulfillment of what in general is nothing but a dream: the hope of describing and explaining "chaotic"nature as the cumulation of many simple steps.

Many fractals have promptly been accepted as works of a new form of art. Some are "representational," while others are totally unreal and abstract. Yet all strike almost everyone in forceful, almost sensual, fashion. The artist, the child, and the "man in the street" never seem to have seen enough, and they had never expected to receive anything of this sort from mathematics. Neither had the mathematician expected his field to interact with art in this way. Eugene Wigner has written about "the unreasonable effectiveness of mathematics in the natural sciences." To this line I have been privileged to add a parallel statement concerning "the unreasonable effectiveness of mathematics as creator of shapes that Man can marvel about and enjoy."

Further reading

The Fractal Geometry of Nature by B.B. Mandelbrot, W. H. Freeman, 1982, was the first comprehensive book on the subject, and remains a basic reference book. Scores of other books have appeared since 1982.

The basic how-to book is *The Science of Fractal Images* (eds. H.-O. Peitgen and D. Saupe, Springer, 1988).

The best known book on iteration is Th*e Beauty of Fractals* by H.-O. Peitgen and P. H. Richter (Springer, 1986).

For other aspects of the mathematics, see *Fractals: Mathematical Foundations and Applications* by K. J. Falconer, (J. Wiley, 1990).

On the concrete uses of fractals, two references are convenient because they are special volumes of widely available periodicals. The first is *Proceedings of the Royal Society of London,* Volume A 423 (May 8, 1989), which was also reprinted as *Fractals in the Natural Sciences,* eds. M. Fleischmann *et al.* (Princeton University Press, 1990). The second is *Physica D.* Volume 38, which was also reprinted as *Fractals in Physics, Essays in Honor of B.B. Mandelbrot on his 65th birthday* (eds. A. Aharony and J. Feder, North Holland, 1989).

On the physics, a standard textbook is *Fractals* by J. Feder (Plenum, 1988).

On some philosophical or social issues, see the Suvol. appended to the 3rd edition of my book, *Les objects fractals* (Flammarion 1989).

Many of my original papers will be reprinted in a multivolume series of *Selecta*; unfortunately, my wish to do them right keeps postponing the publication.

DISCUSSION

HOLTE: We will have a period for discussion and questions and answers at this lecture and after other lectures too. First of all, are there any responses to Dr. Mandelbrot's lecture from the panel?

FEIGENBAUM: Benoit, if I could ask, speaking of poetry and prose—this is a rather flippant question — is it more like music or like noise?

MANDELBROT: For me, music is a form of poetry, and I forgot to say so simply because I felt it was obvious. Analogies can become very dangerous if pursued too far, but I'm glad that you have been taken by the game.

SMALE: Benoit has given us a lot of very nice insights, I think, into this kind of geometry, but I would like to express a little different point of view, which emphasizes something else. I think Benoit said at one point something to the effect that the physical properties reduced to the geometric properties. I think the geometry is very static, and to me the static should best be seen with deeper understanding as flowing from the dynamics Therefore, I would put a dynamical perspective on the understanding of physics above that of a geometrical perspective. From the dynamics, the physical process itself—from the equations, which are time dependent—one can derive some of these fractal geometric pictures with a deeper understanding than by just looking at the pure fractal geometry in its own right. So, for me, the primary emphasis and the deeper physics come from the dynamics rather than the geometry.

MANDELBROT: I see absolutely no conflict between your viewpoint and mine. To study the dynamics of Julia sets, you must study the statics of the Mandelbrot set. In many cases, for example, the shape of the mountains, all one knows well is static. If so, the next step would be to understand the processes that create the mountains. This task is far from complete, but the fractal pictures due to James Bardeen are constructed in successive steps that attempt to make use of what is known of the dynamics in order to represent the statics.

Since very often the geometry of statics is fractal, and the geometry of dynamics is also fractal, fractals do not lose either way.

PRIGOGINE: I would like to say that I'm completely in agreement with what Professor Smale just said. I believe that one of the main points is to relate dynamics to chaos and to fractals. In fact, let me give two examples of what you have said where I think some additional dynamics would be very nice. When we speak about adding some noise, from where is this noise coming? And when you speak about boundary conditions, from where are the boundary conditions coming? Essentially boundary conditions are an empirical concept. If you take hydrodynamics or microscopic physics, you can speak about boundaries. If you speak about dynamics, there are no boundaries. Boundaries are part of the dynamical problem. Therefore, in a sense, I think that your presentation—which was very beautiful, of course—is more about phenomenology which has to be, I would say, made a little deeper by making some connection with dynamical concepts.

MANDELBROT: Two slides which went by very rapidly would have answered your question in a special case. When a fractal aggregate grows, its boundary is continually changed by the dynamics of the generating process. Thus, I agree to what you say. This dynamics consists in little particles aggregating together, but eventually leads to an extraordinary structure. The open mystery is why this structure is fractal.

QUESTION FROM THE AUDIENCE: In the past large mathematical models were used to centralize decisions—for example, in economics. Traditional models have not worked. What does the new science say about prediction, control, and, ultimately, to social responsibility?

MANDELBROT: Your question is very long and complicated. I prefer not to answer the last part.

But I have been greatly interested in economics. In view of your comments, I must emphasize the striking failure of existing economic thought to predict anything about those aspects of the economy on which tests are possible, although data are available in large quantity. For example, many people have long attempted to explain or predict the stock market, but all have failed. When I tackled them in the early 1960s, my approach was very different. It was phenomenological — absolutely, deliberately, and even arrogantly. My goal was "merely" to generate wiggles that people active in the stock market would not be able to distinguish from the wiggles they see in newspapers. This goal was both modest and demanding; I succeeded with the help of a very simple purely random process. One widespread reaction from economists was to challenge me to explain my statistical statics from the economists' dynamics. Disappointingly, those economic theories were not up to the task.

Economics and other more complicated areas borrow a great deal from physics. What they borrow is mostly made of fully developed concepts and theories, such as the concept of equilibrium and the theory of displacement of equilibrium in perfect gases. Next, they try to develop these concepts and themes in rigorous fashion in an economics context. Much less effort is devoted to testing whether economic phenomena really fall into the domain in which those standard physical arguments can conceivably apply.

For example, take continuity. Everyone in economics seemed to assume that prices were a continuous function of time. Well, the evidence is that one comes much closer to reality by assuming prices to be discontinuous functions of time. Incidentally, this discontinuity is not that of quantum physics.

To summarize: I have been very active in economics in the early 1960s — and also again very recently. The reason why my effort in this area has been arrogantly phenomenological is because the more ambitious dynamical study of these things had been an abject failure.

GLEICK: I was interested to hear that that question about complicated economic models was put in the past tense. It continues to be true that a fantastic amount of money and effort is put into enormously complex, many-variable, mostly linear, economic forecasting models. You can read the predictions from these every year in the *Wall Street Journal*. Models attempt to link tens of thousands of variables and relationships—home mortgage interest rates, the ratio of the dollar and the yen, the demand for Sierpinski gaskets—anything you can imagine is built into these models, and the results are often announced to two or three digits of precision. And then, of course, next year, they have to be artificially amended with tens of thousands of ad hoc changes. I think we're only beginning to see an appreciation by some economists of some of the work you've already started to hear described and that you'll hear described as this conference goes on, an appreciation of what can be done with a greater recognition of the essential non-linearity of enormous complex systems like economics.

HOLTE: I have one more question from the audience that we can take time for. When doing mathematical research, do you discover or invent?

MANDELBROT: I certainly feel that I discover. The assertion that eventually became the four-color theorem was discovered long ago—by an amateur. It was not some new thing to be invented, but an existing fact to be discovered. It was there.

The same was true when I sat in front of a terminal, next to an extraordinarily gifted young assistant, to investigate the set which became known as the Mandelbrot set. It was never our feeling that we were inventing anything. This thing was there. Our whole thrust was to discover more about its complication. Its complication was the key to the dynamics of quadratic iteration, which is a dynamical system with particularly simple equations. We tried to discover the so-called static geometry of one set, in order to understand the dynamics of another set.

Let me also mention the work on multifractals which I did in the 1960s and published in 1974. In this instance, the process of discovery occurred on two levels. First of all, I discovered new facts about random singular measures. The key was a mathematical theorem I had learned as a young man, but had always felt would never be used in physics. Hence, it is the study of multifractals that made me discover the real meaning of that theorem. Until then, it had been stated in such an abstract way that I could not see it and appreciate what it had always meant.

Proofs are very often a very different matter. Some are so contrived that they definitely look and feel invented; but the best proofs also have both the look and the feel of discovery.

THE CAUSALITY PRINCIPLE, DETERMINISTIC LAWS AND CHAOS
Heinz-Otto Peitgen

Prediction is difficult, especially of the future.
  —*Niels Bohr*

For many, chaos theory already belongs among the greatest achievements in the natural sciences in this century. Indeed, it can be claimed that very few developments in the natural sciences have awakened so much public interest. Here and there, we even hear of changing images of reality or of a revolution in the natural sciences.

Critics of chaos theory have been asking whether this popularity could perhaps only have something to do with the clever choice of catchy terms or the very human need for a theoretical explanation of chaos. Some have prophesied for it exactly the same quick and pathetic death as that of catastrophe theory, which excited so much attention in the sciences at the end of the 1960s, and then suddenly fell from grace even though its mathematical core is counted as one of the most beautiful constructions and creations. The causes of this demise were diverse and did not only have scientific roots. It can certainly be said that catastrophe theory was severely damaged by the almost messianic claims of some apologists.

Chaos theory, too, is occasionally in danger of being overtaxed by being associated with everything that can be even superficially related to the concept of chaos. Unfortunately, a sometimes extravagant popularization through the media is also contributing to this danger; but at the same time this popularization is also an important opportunity to free areas of mathematics from their intellectual ghetto and to show that mathematics is as alive and important as ever.

But what is it that makes chaos theory so fascinating? What do the supposed changes in the image of reality consist of? To these subjects we would like to pose, and to attempt to answer, some questions regarding the philosophy of nature.

Cause and Effect

The main characteristic of science is its ability to relate cause and effect. On the basis of the laws of gravitation, for example, astronomical events such as eclipses and the appearances of comets can be predicted thousands of years in advance. Other natural phenomena, however, appear to be much more difficult to predict. Although the movements of the atmosphere, for example, obey the laws of physics just as much as the movements of the planets do, weather prediction is still rather problematic.

Ian Stewart, in his article *Chaos: Does God Play Dice?*, makes the following striking comparison:

> Scientists can predicts the tides, so why do they have so much trouble predicting the weather? Accurate tables of the time of high or low tide can be worked out months or even years ahead. Weather forecasts often go wrong within a few days, sometimes even within a few hours. People are so accustomed to this difference that they are not in the least surprised when the promised heat wave turns out to be a blizzard. In contrast, if the tide table predicted a low tide but the beach was under water, there would probably be a riot. Of course the two systems are different. The weather is extremely complex; it involves dozens of such quantities as temperature, air pressure, humidity, wind speed, and cloud cover. Tides are much simpler. Or are they? Tides are perceived to be simpler because they can be easily predicted. In reality, the system that gives rise to tides involves just as many variables — the shape of the coastline, the temperature of the sea, its salinity, its pressure, the waves on its surface, the position of the Sun and Moon, and so on — as that which gives rise to weather. Somehow, however, those variables interact in a regular and

predictable fashion. The tides are a phenomenon of order. Weather, on the other hand, is not. There the variables interact in an irregular and unpredictable way. Weather is, in a word, chaotic.

We speak of the unpredictable aspects of weather just as if we were talking about rolling dice or letting an air balloon loose to observe its erratic path as the air is ejected. Since there is no clear relation between cause and effect, such phenomena are said to have random elements. Yet there was little reason to doubt that precise predictability could, in principle, be achieved. It was assumed that it was only necessary to gather and process greater quantities of more precise information (e.g., through the use of denser networks of weather stations and more powerful computers dedicated solely to weather analysis). Some of the first conclusions of chaos theory, however, have recently altered this viewpoint. Simple deterministic systems with only a few elements can generate random behavior, and that randomness is fundamental; gathering more information does not make it disappear. This fundamental randomness has come to be called chaos.

Deterministic Chaos

An apparent paradox is that chaos is deterministic, generated by fixed rules that do not themselves involve any elements of change. We even speak of deterministic chaos. In principle, the future is completely determined by the past; but in practice small uncertainties, much like minute errors of measurement that enter into calculations, are amplified, with the effect that even though the behavior is predictable in the short term, it is unpredictable over the long run.

The discovery of such behavior is one of the most important achievements of chaos theory. Another is the methodologies which have been designed for a precise scientific evaluation of the presence of chaotic behavior in mathematical models as well as in real phenomena. Using these methodologies, it is now possible, in principle, to estimate the 'predictability horizon' of a system. This is the mathematical, physical, or time-parameter limit within which predictability is ideally possible and beyond which we shall never be able to predict with certainty. It has been established, for example, that the predictability horizon in weather forecasting is not more than about two or three weeks. This means that no matter how many more weather stations are included in the observation, no matter how much more accurately weather data are collected and analyzed, we shall never be able to predict the weather with any degree of numerical accuracy beyond this horizon of time.

Let us set this introductory discussion of what chaos theory is trying to accomplish in a historical perspective. If we look at the development of the sciences on a time-scale on which the efforts of our forebears are visible, we will observe indications of an apparent recapitulation in the present day, even it at a different level. To people during the age of early human history, natural events must have seemed largely to be pure chaos. At first very slowly, then faster and faster, the natural sciences developed, i.e., over the course of thousands of years, the area where chaos reigned seemed to become smaller and smaller. For more and more phenomena, their governing laws were wrung from Nature and their rules were recognized. Simultaneously, mathematics developed hand in hand with the natural sciences, and thus, an understanding of the nature of a phenomenon soon came to also include the discovery of an appropriate mathematization of it. In this way, there was continuous nourishment of the illusion that it was only a matter of time, along with the necessary effort and means, before chaos would be completely banned from human experience.

A landmark accomplishment of tremendous, accelerating effect was made about three hundred years ago with the development of calculus by Sir Isaac Newton (1642-1727) and Gottfried Wilhelm Freiherr von Leibniz (1646-1716). Through the universal mathematical ideas of calculus, the basis was provided with which to apparently successfully model the laws of the movements of planets with as much detail as desired. Calculus also provided the basis for modeling the development of populations, the spread of sound through gases, the conduction of heat in media, the interaction of magnetism and electricity, or even the course of weather events. Also maturing during that time was the secret belief that the terms determinism and predictability were equivalent.

The Laplace Demon

For the era of determinism, which was mathematically grounded in calculus, the "Laplace demon" became the symbol. "If we can imagine a consciousness great enough to know the exact locations and velocities of all the objects in the universe at the present instant, as well as all forces, then there could be no secrets from this consciousness. It could calculate anything about the past or future from the laws of cause and effect."[1]

At its core the deterministic credo means that the universe is comparable to the orderly running of a tremendously precise clock, in which the present state of things is, on the one hand, simply the consequence of its prior state, and on the other hand, the cause of its future state. Present, past, and future are bound together by causal relationships; and according to the views of the determinists, the problem of an exact prognosis is only a matter of the difficulty of recording all the relevant data. The deterministic credo was characteristic of the Newtonian era, which for the natural sciences came to an end, at the latest, through the insights of Werner Heisenberg in the 1927 proclamation of his uncertainty principle[2], but which for other sciences is still considered valid.

---

[1] Pierre Simon de Laplace (1749-1827), a Parisian mathematician and astronomer.

[2] This is also called the indeterminancy principle and states that the position and velocity of an object cannot, even in theory, be exactly measured simultaneously. In fact, the very concept of a concurrence of exact position and exact velocity have no meaning in nature. Ordinary experience, however, provides no evidence of the truth of this principle. It would appear to be easy, for example, to simultaneously measure the position and the velocity of a car; but this is because for objects of ordinary size, the uncertainties implied by this principle are too small to be observable. But the principle becomes really significant for subatomic particles such as electrons.

Strict Causality

Heisenberg wrote: "In the strict formulation of the causality law—'When we know the present precisely'—we can calculate the future" it is not the final clause, but rather the premise, that is false. We cannot know the present in all its determining details.

"Therefore, all perception is a selection from an abundance of possibilities and a limitation of future possibilities.... Because all experiments are subject to the laws of quantum mechanics, and thereby also to the uncertainty principle, the invalidity of the causality law is definitively established through quantum mechanics."

Classical determinism in its fearful strictness had to be given up — a turning point of enormous importance.

Yet the hope in a great victory for determinism persisted undiminished in the first part of this century. This is impressively illustrated in the 1922 book by Lewis F. Richardson entitled *Weather Prediction by Numerical Process*,[3] in which was written: "After so much hard reasoning, may one play with a fantasy? Imagine a large hall like a theater, except that the circles and galleries go right round through the space usually occupied by the stage. The walls of this chamber are painted to form a map of the globe. The ceiling represents the north polar regions, England is the gallery, the tropics in the upper circle, Australia on the dress circle and the antarctic in the pit. A myriad of computers[4] are at work upon the weather of the part of the map where each sits, but each computer attends only to one equation or part of an equation. The work of each region is coordinated by an official of higher rank. Numerous little 'night signs' display the instantaneous values so that neighboring computers can read them.... From the floor of the pit a tall pillar rises to half the height of the hall. It carries a large pulpit on its top. In this sits the man in charge of the whole theater; he is surrounded by several

---

[3] Dover Publications, New York, 1965. First published by Cambridge University Press, London, 1922. This book is still considered on of the most important works on

[4] Richardson uses the word computer here to mean a person who computes.

assistants and messengers. In this respect he is like the conductor of an orchestra in which the instruments are slide-rules and calculating machines. But instead of waving a baton he turns a beam of rosy light upon any region that is running ahead of the rest, and a beam of blue light upon those who are behindhand."

In his book, Richardson first laid down the basis for numerical weather forecasting and then reported on his own initial practical experience with calculation experiments. According to Richardson, the calculations were so long and complex that only by using a 'weather forecasting center' such as the one he fantasized was forecasting conceivable.

Then about the middle of the 1940s, the great John von Neumann actually began to construct the first electronic computer, ENIAC, in order to further pursue Richardson's prophetic program, among others. It was soon recognized, however, that Richardson's only mediocre practical success was not simply attributable to his equipment's lack of calculating capacity, but also to the fact that the space and time increments used in his work had not met a computational stability criterion (Courant-Friedrichs-Lewy Criterion), which was discovered only later. With the appropriate corrections, further attempts were soon underway with progressively bigger and faster computers to make Richardson's dream a reality. This development has been uninterrupted since the 1950s, and it has bestowed truly gigantic "weather theaters" upon us.

Weak Causality

Indeed, the history of numerical weather forecasting illustrates better than anything else the undiminished belief in a deterministic (viz., predictable) world; for in reality, Heisenberg's uncertainty principle did not at all mean the end of determinism. It only modified it, because scientists had never really taken Laplace's credo completely seriously — as is usual with creeds. The most carefully conducted experiment is, after all, never completely isolated from the influences of the surrounding world, and the state of a system is never precisely known at any point in time. The absolute mathematical precision that Laplace presupposed is not physically realizable;

minute imprecision is, as a matter of principle, always present. What scientists actually believed was this: from approximately the same causes follow approximately the same effects— in nature as well as in any good experiment. And this is indeed the case, especially over short time spans. If this were not so, we would not be able to ascertain any natural laws, nor could we build any functioning machines.

The Butterfly Effect

But this apparently very plausible assumption is not universally true. And what is more, it does not do justice to the typical course of natural processes over long periods of time. Around 1960, Ed Lorenz discovered this deficiency in the models used for numerical weather forecasting; it was he who coined the term "butterfly effect." His description of deterministic chaos goes like this:[5] chaos occurs when the error propagation, seen as a signal in a time process, grows to the same size or scale as the original signal.

Thus, Heisenberg's response to deterministic thinking was also incomplete. He concluded that the strong causality principle is wrong because its presumptions are erroneous. Lorenz has now shown that the conclusions are also wrong. Natural laws, and for that matter determinism, does not exclude the possibility of chaos. In other words, determinism and predictability are not equivalent. And what is an even more surprising finding of recent chaos theory is the discovery that these effects are observable in many systems which are much simpler than the "weather." In fact, they can be observed in very simple feedback systems, even as simple as the quadratic interator $x \longrightarrow ax(1-x)$.

Moreover, chaos and order (i.e., the causality principle) can be observed in juxtaposition within the same system. There may be a linear progression of errors characterizing a deterministic system which is governed by the causality principle, while (in the same system) there can also be

---

[5] See Peitgen, H.-O, Jürgens, H., Saupe, D., and Zahlten, C., Fractals — An Animated Discussion, Video film, Freeman 1990. Also appeared in German as *Fraktale in Filmen und Gesprächen,* Spektrum der Wissenschaften Videothek, Heidelberg, 1990.

THE CAUSALITY PRINCIPLE, DETERMINISTIC LAWS AND CHAOS 43

an exponential progression of errors (i.e., the butterfly effect) indicating that the causality principle breaks down.

So twentieth century science has undermined both the premise and conclusion of the causality principle. In other words, one of the significant lessons coming out of chaos theory is that the validity of the causality principle is narrowed by the uncertainty principle from one end as well as by the intrinsic instability properties of the underlying natural laws, i.e., chaos, from the other end.

THE TRANSITION TO CHAOS
Mitchell J. Feigenbaum

The study of chaos is a study that is part of a larger program of study of so-called "strongly" nonlinear systems. Within the context of the discipline of physics, the exemplar of such a system is a fluid in turbulent motion. If chaos is not exactly the study of fluid turbulence, nevertheless, the image of turbulent, erratic motion serves as a powerful icon to remind a physicist of the sorts of problems he would ultimately like to comprehend. As for all good icons, while a vague impression of what one wants to know is sensibly clear, a precise delineation of many of these quests is not so readily available. In a state of ignorance, the most poignantly insightful questions are not yet ripe for formulation. Of course, this comment remains true despite the fact that, for technical exigencies, there are definite questions that one desperately wants the answers to.

Fluid turbulence indeed presents us with highly erratic and only partially predictable phenomena. Historically, since Laplace, say, physical scientists have turned to the framework of statistical methods when presented with problems that concern the mutual behaviors of innumerably large numbers of pieces. If for no other reason, one does so to reduce the number of details that one must measure, specify, compute, whatever. Thus, it is easier to say 43% of the population voted for X rather than to offer the roster of the behavior of each of millions of voters. Just so, it is easier to specify how many gas molecules there are in an easily measurable volume than to write out the list of where and how fast each one is. This idea is altogether reasonable if not even the most desirable one. However, if one is to work out a theory of these things, so that a prediction might be rendered, then, as in all matters of statistics, one must determine a so-called distribution function. This means a theoretical prediction of just how often out of uncountably many elections, etc., it is expected that each value of this average voter response occurs. For the voter question and the density-of-a-gas question, there is just one number to determine. For the problem of fluid turbulence, even in this statistical quest, one must ask a much richer question. For example, how often do we see eddies of each size rotating at such and such a rate?

For the problem of voters I don't have any serious idea of how to theoretically determine this requisite distribution; nor with good frequency, do the polls succeed in measuring it. After all, it might not exist, in the sense that it rapidly and significantly varies from day to day. However, since physicists have long known quite reliably the laws of fluids—that is, the rules that allow you to deduce what each bit of fluid will do later if you know what they all do now, there might be a way of doing so. Indeed, the main idea of the branch of physics called statistical mechanics is rooted in the belief that one knows in advance how to do this. The idea is, basically, that each possible detailed configuration occurs with equal likelihood. Indeed the word "chaos" first entered physics in the last century in Maxwell's phrase, "state of molecular chaos," to loosely mean this. Statistical mechanics—especially in its quantum mechanical form—works very well indeed, and provides us with some of our most wonderful knowledge. However, altogether regrettably, in the context of fluid turbulence, it has persisted for the last century to roundly fail. It turns out to be a question of truly deducing from the known laws of microscopic motion of fluids what this rule of distribution must be, because the easy guess of "everything is as random as possible" simply doesn't work. And when that guess doesn't work, there exists as of today no methodology to provide it. Moreover, if in our present state of knowledge we should be forced to appraise the situation, then we would guess that an extraordinarily complicated distribution is required to account for the phenomena. Should it be fractal in nature, then it is fractal of the most perverse sort. And the worst part is that we really don't possess the mathematical power to generally say what class of objects it might be sought among. Remember, we're not looking for a perfectly good quick fix: If we are serious in seeking understanding of the analytical description of Nature, then we demand much more. When the subject of chaos and a part of that larger problem called strongly nonlinear physics shall have been deemed penetrated, we shall know thoroughly how to respond to such questions, and readily imagine intuitively what the answers look like. To date, we can now compellingly do so only for much simpler problems—and have come to possess that capability only within the last decade.

As I said earlier, I don't necessarily care especially about turbulence. Rather, it serves as an icon representing a genre of problems. I was trained as a theoretical high-energy physicist, and grew deeply troubled that no methods save for that of successive improvements, so-called perturbation methods, existed. Apart from the brilliant effort of Ken Wilson, in his version of the renormalization group, that circumstance is unchanged. Knowing the microcosmic laws of how things move—such schemes are called "dynamical systems"—still leaves us almost altogether in the dark as to their larger consequences. Are the theories no good, or is it that we just can't determine what they contain? At the moment it's impossible to say. From high-energy physics to fluid physics our inherited ways of thinking mathematically simply fail to serve us. In a way, if perhaps modest, the questions tackled in the effort to comprehend what is now called chaos have faced these questions of methodology head on.

Let me now backtrack and discuss nonlinearity. This means, first, linearity. Linearity means that the rule that determines what a piece of a system is going to do next is not influenced by what it is doing now. More precisely, this is intended in a differential or incremental sense; for a linear spring, the increase of its tension is proportional to the increment whereby it is stretched, with the ratio of these increases exactly independent of how much it has already been stretched. Such a spring can be stretched arbitrarily far, and in particular will never snap or break. Accordingly, no real spring is linear.

The mathematics of linear objects is particularly felicitous. As it happens, linear objects enjoy an identical, simple geometry. The simplicity of this geometry always allows a relatively easy mental image to capture the essence of a problem, with the technicality, growing with the number of parts, basically a detail, until the parts become infinite in number, although often then too, precise answers can be readily determined.

The historical prejudice against nonlinear problems is that no so simple nor universal geometry usually exists. Until recently, the general scientific perception was that a certain nonlinear equation characterized some particular problem. If the specific problem was sufficiently interesting or demanding of resolution, then, perhaps, particular methods could be created for it while, however, it was well understood that the travail would probably be of no avail in other contexts.

Indeed only one method was well understood and universally learned, the so-called perturbation method. If a linear problem is viewed through distorting lenses, it qualitatively will do the same thing; if it repeated every five seconds, that is, the motion was periodic with a period of five seconds, it would persist to appear so seen through the lenses. Nevertheless, it would now no longer appear to exhibit equal tension increments for the equal elongations: After all, the tension is measurably unchanged by distorting lenses, whereas all spatial measurements are. That is, the device of distorting lenses turns a linear problem into a nonlinear one. The method of perturbation basically works only for nonlinear problems that are distorted versions of linear ones. And so, this uniquely well-learned method is of no avail in matters that aren't merely distortions of linear ones.

Chaos is absent in distorted linear problems. Chaos and other such phenomena that are qualitatively absent in linear problems are what we call strongly nonlinear phenomena. It is this failure to subscribe to the spectrum of configurations allowed by distorting a simple geometry that renders these problems anywhere from hard in the extreme to impenetrable. How does one ever start to intelligently describe an awkward new geometry? This question is, for example, intended to be loosely akin to the question of how one should describe the geometry of the surface of the Earth, not through our abstracted perceptual apparatus that allows us to visualize it immersed within a vastly larger three dimensional setting, but rather intrinsically, forbidding this use of imagination. The solution of this question, first by Gauss and then extended to arbitrary dimensions by Riemann is, as many of you must know, at the center of the way of thinking of Einstein's General Theory of Relativity, our theory of gravity. What is to be the geometry of the object that describes the turbulent fluid's distribution function? Are there intrinsic geometries that describe various chaotic motions, that serve as a unifying way of viewing these disparate nonlinear problems, as kindred? I ask the question because I know the answer to be in the affirmative in certain broad circumstances. The moment this is accepted, then strongly nonlinear problems appear no longer as each one its own case, but rather coordinated and suitable for theorizing upon as their own abstract entity. This promotion from the detailed specific to the membership in a significant general class is one of the triumphs of the study of chaos in the last decade or two.

An even stronger notion than this generality of shared qualitative geometry is the notion of universality, which means no less than that shared geometry is not only one of a qualitative similarity but also one of true quantitative identicality. After what has been, if you will, a long preamble, the fact that strongly nonlinear problems, with surprising frequency, can share a quantitatively identical geometry is what I shall pursue for the rest of this talk, and constitutes what is termed universality in the transition to chaos.

In a qualitative way of thinking, universality can be seen to be not so surprising. There are two arguments to support this. The first part has simply to do with nonlinearity. Just as a linear object has a constant coefficient of proportionality between, for example, its tension and its expansion, a similar, but nonlinear version, has an <u>effective</u> coefficient <u>dependent</u> upon its extension. So, consider two completely different nonlinear systems. By adjusting things correctly it is not inconceivable that the effective coefficients of each part of each of the two systems can be set the same so that then their behaviors could, at least initially, be identical. That is, by setting some numerical constants (properties, so to speak, that specify the environment, mathematically called "parameters") <u>and</u> the actual behaviors of these two systems, it is possible that they can do the identical thing. For a linear problem this is ostensibly true; for systems with the same number of parts and mutual connections, a freedom to adjust all the parameters allows one to be adjusted to be identical (truly) to the other. But, for many pieces, this is many adjustments. For a nonlinear system, adjusting a small number of parameters can be compensated, in this quest for identical behavior, by an adjustment of the momentary position of its pieces. But then it must be that not all motions can be so duplicated between systems.

Thus, the first part of the argument is that nonlinearity confers a certain flexibility upon the adaptability of an object to desirable behavior.

Nevertheless, should the precise adjustment of too many specific and subtle details be required in order to achieve a certain universal behavior, then the idea would be pedantic at best.

However, there is a second, more potent, argument, a paraphrasing of Leibniz in "The Monadology," which can render this first argument potent. Let us contemplate that the motion we intend to determine to be universal over nonlinear systems has arisen by the successive imposition of more and more qualitative constraints. Should this growingly large host of impositions prove to be generally amenable to such systems (this is the hard and *a priori* neither obvious nor reasonable part of discussion), then we shall ultimately discover these disparate systems to all be identically constrained by an infinite number of qualitative and, if you will, self-consistent, requirements. Now, following Leibniz, we ask, "In how many precise, or quantitative, ways can this situation be tenable?" And we respond, following Leibniz, by asserting, "In precisely one possible uniquely determined way."

This is the best verbalization I know how to offer to explain why such a universal behavior is possible. Both mathematics and physical experimentation confirm its rectitude perfectly. But it is perhaps difficult to have you realize how extraordinary this result appeared given the backdrop of physical and mathematical thinking in 1976 when it first appeared together with its full conceptual analysis. As anecdotal evidence, I had been directed to expound these results to one of the great mathematicians who is renowned for his results on dynamical systems. I spoke with him at the very end of 1976. I kept trying to tell him that there was a complete quantitative universality to these phenomena, and he equally often understood me to have duplicated some known qualitative results. Finally he said, "You mean to tell me these are metrical results?" (Metrical is a mathematical code word that means quantitative.) And I said, "Yes." "Well, then you're wrong!" he asserted, and turned his back on me to terminate the conversation.

Anecdote aside, what is remarkable about all this? First of all, an easy piece of methodological insight. As practitioners of a truly analytical science, we physicists were trained to know that qualitative explanations are insufficient to base truth upon. Quite to the contrary, it is regarded to be at the heart of the so-called "scientific method" that ever more precise measurements will discriminate between rival quantitative theories to ultimately select out one as the correct encoding of the qualitative content. (Thus, think of geocentric versus heliocentric planetary theories, both qualitatively explaining the retrograde motions of the planets.) Here the method is turned on its head. Qualitatively similar phenomena, independently of any other ideational input, must ineluctably lead to the measurably identical quantitative result. Whence the total phenomenological support for this mighty "scientific method"?

Secondly, a new principle of "economy" immediately emerges. Why put out Herculean efforts to calculate the consequences of some particular and highly difficult encoding of physical laws, when anything else, however trivial, possessing the same qualitative properties will yield exactly the same predictions and results? And this is all the more satisfying, since one doesn't even know the exact equations that describe various of these phenomena, fluid phenomena in particular. These phenomena have nothing to do, whatsoever, with the detailed, particular, microscopic laws that happen to be at play. This aspect—that is, of substituting easy problems for hard ones with no penalty—has been, as a way of thinking and performing research, the prominent fruit of the recognition of universality. When can it work? Well, in complicated interactions of scores of chemical species, in laser phenomena, in solid state phenomena, in, at least partially, biological rhythmic phenomena such as apneas and arrhythmias, in fluids, and, of course, in mathematics.

But now, as I move towards the end of this claim for virtue, and also towards the end of this talk, let me discuss "chaos" a bit more per se and revisit my opening "preamble." Much of chaos as a science is connected with the notion of "sensitive dependence on initial conditions." Technically, we scientists term as "chaotic" those nonrandom complicated motions that exhibit a very rapid growth of errors that, despite perfect determinism, inhibits any pragmatic ability to render accurate long-term prediction. While nomenclaturally speaking, this is perforce true, I personally am not most intrigued nor concerned with this facet of my subject.

I've never told you what the "transition to chaos" means, but you can readily guess from the verbiage that it's something that starts off not being chaotic, ends up being so, and hence somehow passes from one to the next. The most important fact is that there is a discernibly precise "moment," with a corresponding behavior, which is neither chaotic nor nonchaotic, at which this transition occurs. Yes, errors do grow, but only in a marginally predictable, rather than in an unpredictable fashion. In this state of marginal predictability inheres embryonically all the seeds of the chaotic behavior to come. That is, this transitional point, the legitimate child of universality, without full-fledged sensitive dependence upon initial conditions, knows fully how to dictate to its progeny in turn how this latter phenomenon must unfold. For a certain range of possible behaviors of strongly nonlinear systems, this range surrounding the transition to chaos, the information obtained just at the transition point fully organizes the spectrum of behaviors that these chaotic systems can exhibit.

Now what is it that turns out to be universal? The answer, mostly, is a precise qualitative determination of the intrinsic geometry of the space upon which this marginal chaotic motion lives together with the full knowledge of how, in the course of time, this space is explored. Indeed, it was from the analysis of universality at the transition to chaos that we came to recognize the precise mathematical object that fully furnishes the intrinsic geometry of these sort of spaces. This object, a so-called scaling function, together with the mathematically precise delineation of universality, constitutes one of the major results of the study of chaos. Granted the broad range of objects that can be termed fractal, these geometries are fractal—but not the heuristic sort of "dragons," "carpets," "snowflakes," etc. Rather, these are structures which are elaborated upon at smaller and smaller scales, differently at each point of the object, and so are infinitely more complicated than the above heuristic objects. There is, in more than just a way of speaking, a geometry of these dynamically created objects, and that geometry requires a scaling function to fully elucidate it. Many of you are aware of the existence of a certain object called the "Mandelbrot set." Virtually none of you, though, even having simulated it on your own computers, are aware that its ubiquitous existence in those sufficiently smooth contexts in which it appears, is the consequence of universality at the transition to chaos. Every one of its details is implicit in those embryonic seeds I have mentioned before. Thus, as but simply the most elementary consequence of this deep universal geometry, in its gross organization we notice a set of discs—the largest the main cardiod—one abutting upon the next and of rapidly diminishing radii. How rapidly do they diminish in size? In fact, each one is δ times smaller than its predecessor, with δ a universal constant, approximately equal to 4.6692016..., the best known of the constants that characterize universality at the transition to chaos.[*]

I have now come around full circle to my introductory comments. We have, in the last decade, succeeded in coming to know many of the correct ideas and their mathematical language in regard to the question, "What is the nature of the objects upon which we see our statistical distributions?" "Dimension" is a mathematical word possessing a quite broad range of technical connotations. Thus, the theory of universality is erected in a very low (that is, one- or two-) dimensional setting. However the information discussed is of an infinite-dimensional character. The physical phenomena exhibiting these behaviors can appear, for example, in the physical three-dimensional space of human experience, with the number of interacting, cooperating pieces that comprise the system investigated—also a statement of its dimension—either merely a few or an infinitude. Nevertheless, our understanding to date is of what must be admitted to be a relatively simple set of phenomena—relatively simple in comparison to the swirling and shattering complexity of fluid motions at the foot of a waterfall, phenomena that still loom large in front of us, deeply impressing upon us how much lies all undiscovered before us.

---

[*] This value is known as the Feigenbaum constant. —Ed.

TIME, DYNAMICS AND CHAOS
Integrating Poincaré's "Non-Integrable Systems"

Ilya Prigogine

I. Questioning Time

Time has always haunted man. Time is indeed our fundamental existential dimension. It has fascinated philosophers as well as scientists. It has often been stated that science has solved the problem of time. Is this really true? Indeed, a fundamental property of the basic equations of physics, be it classical physics or quantum physics, is time reversibility. We may, in these equations, replace t by -t without changing the form of these equations. In contrast, in the macroscopic world, we deal with irreversible processes: +t and -t do not play the same role. There exists an "arrow of time." We come, therefore, to the strange conclusions that in the microscopic dynamic world, there would be no natural time ordering in contrast to what happens in the macroscopic world. For example, if we consider two positions of a pendulum, as represented in Figure 1, we cannot say which position comes earlier. With classical dynamics, time has lost its direction.

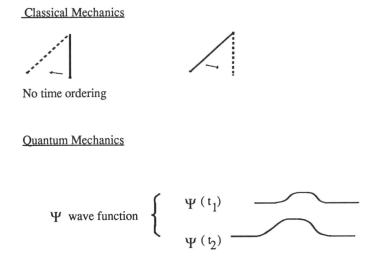

Figure 1: No time ordering in classical or quantum dynamics.

Similarly, in quantum mechanics, we cannot speak about "older" or "younger" wave functions. But can this be the whole story? How can time emerge from a time-reversible world? This conflict has become quite evident since the formulation by Clausius in 1865 of the well-known second law of thermodynamics. Clausius stated, "The entropy of the universe is increasing."

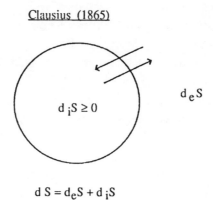

Figure 3. Clausius formulation of the second law of thermodynamics

Figure 3. Clausius formulation of the second law of thermodynamics
This was the birth of evolutionary cosmology. For every isolated system, entropy can only increase. Entropy expresses, as Eddington used to say, "the arrow of time."

The formulation of thermodynamics was the result of the work of engineers and physical chemists. The great mathematicians and physicists of this time considered it at best a useful practical tool, but without any functional significance. The first to ask the question of the relation between entropy and the microscopic equations of motion was Boltzmann. Boltzmann was one of the main founders of kinetic theory. He tried to explain the increase of entropy as the result of molecular collisions leading to molecular disorder (the Maxwell velocity distribution law). Boltzmann's approach is still of great importance today, as it leads, for dilute gases, to results that are in excellent agreement with experiment. Still, Boltzmann was defeated, as people were quick to point out to him that his results clashed with dynamic time reversibility (see, e.g., [1]). Boltzmann was like a man in love with two women. He could not choose between his conviction that time-irreversible evolution was an essential aspect of nature, and his confidence in the classical equations of motion which seem to prevent the existence of a privileged direction of time. I cannot go into details about this question, but let me stress that one of the aims of this lecture is to show that Boltzmann was right, but this involves quite recent results in which modern chaos theory plays an essential role. For Boltzmann's generation, as well as for the generations that followed, the conclusion of this debate was that the arrow of time was not in nature, but in our mind. Einstein's saying, time (as irreversibility) is an "illusion," is well known [1].

I have always found it curious that this conclusion did not trigger a crisis in science. How can we deny the existence of a privileged direction of time? As Popper wrote, "This would brand unidirectional change as an illusion. This would make our world an illusion and with it, all our attempts to find more about our world."

The ambition of classical science was to describe the behavior of nature in terms of universal, time-reversible laws. It is interesting to reflect on the relation between this ambition and the theological concepts that prevailed in the 17th century. For God, there is, of course, no distinction between past, present, and future. Is science not bringing us closer to God's view of the universe? [2] This ambition of classical science was never realized. Often, science seemed close to this goal, and every time, something failed. This gives a dramatic form to the history of western science. As you know, quantum mechanics is based on Schrödinger's equation, which is time reversible, but one had to introduce the measurement process and with it to attribute a fundamental role to the observer to obtain a consistent description. Einstein started general relativity as a geometrical, "timeless" theory, only to discover the need for some initial singularity or instability to obtain a consistent description of the cosmological evolution of our universe. To describe nature, we need both laws and events, and this, in turn, implies a temporal element, which was missing in the traditional presentation of dynamics, including quantum theory and relativity.

Last year's Nobel Conference in Minnesota had the provocative title, "The End of Science?" I don't believe we can speak about the end of science, but indeed we come to the end of a certain form of rationality associated to the classical ideology of science. As I want to show here, in the building up of this new scientific rationality, non-equilibrium physics and "chaos" certainly will play an essential role.

II. The Time Paradox Builds Up

The 20th century is characterized by the discovery of quite unexpected features in which the arrow of time is essential. Examples are the discovery of unstable elementary particles and of evolutionary cosmology. I would like, however, first to emphasize in this lecture processes involving a macroscopic scale such as those studied in non-equilibrium physics.

A first remark: Contrary to what Boltzmann believed, irreversibility plays a <u>constructive</u> role. It not only is involved in processes leading to disorder, but also can lead to order. This already appears in very simple examples such as those presented in Figure 3. Consider two boxes containing two components, say hydrogen and nitrogen. If the boxes were at the same temperature, the proportion of these two components would be the same in the two compartments. If, on the contrary, we establish a temperature difference, we observe that the concentration of one of the components, say hydrogen, becomes larger in the compartment that is at the higher temperature.

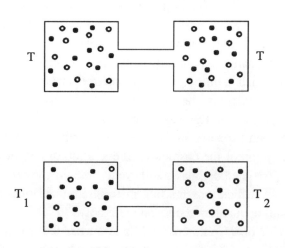

Figure 3. Thermal diffusion experiment (see text)

The disorder associated with the flow of heat is used in this experiment to create "order." This is quite characteristic. Irreversibility leads both to order and disorder.

A striking example is the case of chemical oscillations. Suppose we have a chemical reaction that may transform "red" molecules into "blue" ones and vice versa. It has been shown both theoretically and experimentally that, far from equilibrium, such a reaction may present a time-periodic behavior. The reaction vessel becomes in succession red, then blue, and so on. Let me emphasize how unexpected the appearance of chemical coherence is. We usually imagine chemical reactions as the result of random collisions between molecules. Obviously, this cannot be the case far from equilibrium. We need long-range correlations to produce chemical oscillations. When we push such systems further away from equilibrium, the oscillations may become quite irregular in time. One then speaks about "dissipative chaos"; however, I shall not go into more details about this subject, which is treated adequately in many texts [3,4].

What is important is that irreversibility leads to new spacetime structures (which I have called "dissipative structures"), and which are essential for the understanding of the world around us. Therefore, irreversibility is "real," it cannot be in our mind, and we have to incorporate it in one way or another in the frame of microscopic dynamics. Recently,

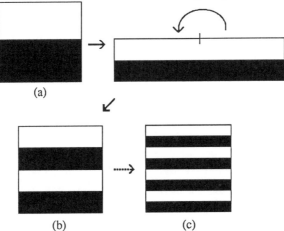

Figure 4. Baker transformation (see text).

there have been many monographs dealing with this problem, but I mention here only the excellent introduction due to Peter Coveney and Roger Highfield, *The Arrow of Time* [5]. In this book, they called this problem "time's greatest mystery."

But how to go beyond this paradox?

In the work done by my colleagues and me, we have followed the idea that the arrow of time must be associated with <u>dynamical instability</u>. Let me first present a very simple example of an unstable dynamic system, the so-called Baker transformation. (See Figure 4.) We consider a square. We squash it and put the right part on the top of the left as seen Figure 4. This leads to a progressive fragmentation of the surface of the square. This is obviously an unstable dynamic system, as two points as close as one wants will finally show up in distant stripes. Such a system can be characterized by a Lyapounov exponent:

$$(\delta x)_t = (\delta x)_0 \exp(\lambda t).$$

The distance $(\delta x)_t$ between two neighboring trajectories increases exponentially with time. The coefficient, λ, is called the Lyapounov exponent, and is, in the case of the Baker transformation, equal to lg 2.

The existence of a positive Lyapounov exponent is characteristic of "chaotic" systems. There exists, then, a temporal horizon beyond which the concept of trajectory fails, and a probabilistic description is to be used. This is all well known. What I want to emphasize, however, is that in addition to the kinematic time t, we can introduce for such systems a second, internal time, T, which measures the number of shifts required to produce a given partition. For example, starting from the state [a] of Figure 4, we need two shifts to obtain state [c]. As has been shown by our group, especially in the work of Prof. Misra [6], this internal time is represented by an operator that leads to a non-commutative algebra very much like one we use in quantum mechanics. Once you have the internal time, it is easy to construct an entropy and therefore to associate to the Baker transformation an arrow of time.

It is important to notice that the internal time refers to a global property as expressed by the partitions of the square. It is a "topological" property It is only in considering the square as a whole that you can associate to it a given internal time, or an "age." It is like when you look at some person. The age you will attribute to him does not depend on a specific detail of his body, but results from a global judgment. A detailed presentation of the Baker transformation and its relation to the arrow of time can be found in my book with Prof. Nicolis [4]. However, the Baker transformation corresponds to a highly idealized situation, and it is not clear on this basis why the arrow of time would be so prevalent in nature, as testified by the universal validity of the second law of thermodynamics. That is the problem to which I want to turn now.

III. Poincaré's Theorem and the Science of Chaos—Large Poincaré Systems

In 1889, Poincaré asked a fundamental question [7,8]. I should mention that his question was not formulated in these terms, but this is the formulation I shall use for the sake of the discussion. Poincaré asked if the physical universe is isomorphic to a system of non-interacting units. As is well known, the energy (the "Hamiltonian" H) is generally formed by the sum of two terms, the kinetic energy of the units involved and the potential energy corresponding to their interactions. Therefore, Poincaré's question was, "Can we eliminate the interactions?"

This is indeed a very important question. If Poincaré's answer had been yes, there could be no coherence in the universe. There would be no life, and no Nobel Conferences. So it is very fortunate that he proved that you cannot, in general, eliminate interactions; moreover, he gave the reason for this result. The reason is the existence of resonances between the various units.

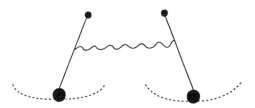

Figure 5. Resonance between coupled oscillators.

Everybody's familiar with the idea of resonance. This is the way the children learn to swing. Let's formulate more precisely Poincaré's question. We start with a Hamiltonian of the form,

$$H = H(p,q) \tag{3.1}$$

where p,q are the momenta and the coordinates. We then ask if we can reduce it to the form

$$H = H(J) \tag{3.2}$$

where J are the new momenta (the so-called action variables). In this form the Hamiltonian depends only on the momenta. To perform the transformation from (3.1) to (3.2) Poincaré considered the class of transformations which conserve the structure of the Hamiltonian theory (so-called canonical or unitary transformations). More precisely, Poincaré considered Hamiltonians of the form

$$H = H_0(J) + \lambda V(J,\alpha) \tag{3.3}$$

where λ is the coupling constant and V is the potential energy which depends both on the J and the coordinates α (called the angle variables). For two degrees of freedom the potential can be expanded in a Fourier series

$$V(J_1, J_2, \alpha_1, \alpha_2) = \sum_{n_1,n_2} V_{n_1,n_2}(J_1, J_2) e^{i(n_1\alpha_1 + n_2\alpha_2)} \tag{3.4}$$

where n_1, n_2 are integers. The application of perturbation techniques leads then to expressions of the form

$$\frac{V_{n_1,n_2}}{n_1\omega_1 + n_2\omega_2} \tag{3.5}$$

with the frequencies ω_i defined as $\omega_i = \partial H_0/\partial J_i$. Here we see the dangerous role of resonances (or "small denominators"):

$$n_1\omega_1 + n_2\omega_2 = 0. \tag{3.6}$$

Obviously we expect difficulties when (3.6) vanishes while the numerator in (3.5) does not. This has been called by Poincaré [7] the "fundamental difficulty of dynamics." We come in this way to Poincaré's classification of dynamical systems [7,8,9]. If there are "enough" resonances, the system is "non-integrable."

Decisive progress in our understanding of the role of the resonances was achieved in the 1950s by Kolmogorov, Arnold, and Moser (the so-called KAM theory). (See, e.g., [3].) They have shown that if the coupling constant in (3.3), λ, is small enough (and also other conditions which I shall not discuss here are satisfied), "most trajectories remain periodic as in integrable systems. This is not astonishing. Formula (3.6) can be written as

$$\frac{\omega_1}{\omega_2} = -\frac{n_2}{n_1},$$

a rational number. Now rationals are "rare" as compared to irrationals. However, whatever the value of the coupling constant λ, there appear now, in addition, random trajectories characterized by a positive Lyapounov exponent and therefore by "chaos." This is indeed a fundamental result, since it is quite unexpected to find randomness at the heart of dynamics, which was always considered to be the stronghold of deterministic description. However, it should be emphasized that the KAM theory has not solved the problem of the integration of Poincaré's non-integrable systems. The statement by Arnold that even dynamical systems with only two degrees of freedom lie beyond our present mathematics has been widely quoted.

But there is a class of dynamical systems we call large Poincaré systems [LPS], for which we may indeed eliminate entirely Poincaré's divergences and therefore indeed "integrate" a class of Poincaré's "non-integrable" systems. The result is the outcome of years of research with my colleagues in Brussels and Austin [9]; however, it is only recently that the problem of the integration of large Poincaré systems has been solved. I want to acknowledge from the start the fundamental contributions of Tomio Petrosky as well as of Hiroshi Hasegawa and Suichi Tasaki [10-14].

First, what is a large Poincaré system? It is a system with a "continuous" spectrum. For example, the Fourier series in formula (3.4) now has to be replaced by a Fourier integral. The resonance conditions take then a new form. The resonance conditions for a small system with an arbitrary number of degrees of freedom are (see (3.4))

$$n_1\omega_1 + n_2\omega_2 + n_3\omega_3 + ... = 0 \qquad (3.7)$$

where the n_i are integers. As mentioned, the resonance conditions express the existence of rational relations between frequencies. For large Poincaré systems, condition (3.4) has to be replaced by

$$k_1\omega_1 + k_2\omega_2 + k_3\omega_3 + ... = 0 \qquad (3.8)$$

where the k_i are <u>real</u> numbers. Now resonances are "everywhere." The situation becomes similar to that in the Baker transformations, where also almost all motions are random motions. Moreover, large Poincaré systems are characterized by interactions involving integrations of resonances.

Before I consider examples, let me emphasize that the idea of large Poincaré systems remains meaningful in quantum mechanics. The frequencies ω_i then become energy levels. For small systems, the resonance condition (3.7) would correspond to accidental "degeneracies." But for large Poincaré systems, we have a continuous spectrum and the situation becomes quite similar to that in classical mechanics. Large Poincaré systems have a surprising generality. We meet them everywhere both in classical and in quantum physics.

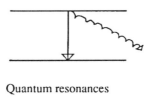

Quantum resonances

Intergration over resonances

(~ Fermi golden rule)

Figure 6. Quantum Transition

Let me present two examples. The interaction between matter and electromagnetic fields leads to the emission of radiation (see Figure 6).The lifetime of the excited states is given to a first approximation by what physicists call Fermi's golden rule, which involves an integration over resonances

$$\int dk \, |V_k|^2 \, \delta(\omega_k - \omega_1) \tag{3.9}$$

where ω_k are frequencies associated to the radiation and ω_1 is the energy level associated to the unstable state. This is an example of integration of resonances. All many-body systems involving "collisions" are LPS (see Figure 7), as collisions also involve resonances (see Section V).

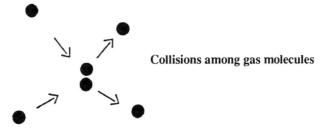

Figure 7. Collisions (text)

Large Poincaré systems are not integrable in the usual sense because of the Poincaré resonances, but what we want to point out in this lecture is that we can integrate them through new methods, eliminating all Poincaré divergencies. This leads to a new "global" formulation of dynamics (classical or quantum). As we deal here with chaotic systems, we may expect new features in this formulation of dynamics. Indeed, we shall find, as compared with the dynamics of integrable systems, an increased role of randomness, and above all a breaking of time symmetry and therefore the emergence of irreversibility at the heart of this new dynamics. In a sense, we invert the usual formulation of the time paradox. The usual attempt was to try to deduce the arrow of time from a dynamics based on time-reversible equations. In contrast, we now generalize dynamics to include irreversibility.

We may summarize the situation as follows. Of course, KAM theory is only valid for classical systems; as mentioned, for quantum systems, we have to consider "large systems," (with continuous spectrum).

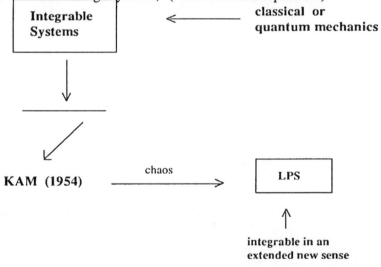

Diagram I

On top we have the class of integrable systems (classical or quantum). This is the main field explored by dynamics. The basic structure of the dynamical integrable systems is expressed by celebrated laws (or principles) such as the action principle, which states that the trajectory is such that some functional (the action) is minimum. This structure has been the starting point both for quantum mechanics and general relativity.

But Poincaré's theorem limits the class of integrable systems. The basic question is then: what happens next? How does nature solve the problem of the small denominators? After all, computer calculations do not lead to infinities!

As mentioned, a first step was the KAM theory. The physical effect of resonances is the appearance of random motion. For LPS almost all motions are random. This is the general case as realized, i.e., in scattering. In simple situations, such as the Friedrichs model studied in section IV, Poncaré's divergences lead to irreversibility, but the concept of wave functions remains valid (in a "rigged Hilbert space"). The remarkable fact is that we then again can "integrate" the equations of motion. But now the structure of dynamics becomes radically different from that of integrable systems. It is fascinating that we now have to deviate from the structure inherent in the dynamical scheme associated with the classical tradition.

IV. Poincaré's Theorem and the Quantum Mechanical Eigenvalue Problem

The first example I want to consider refers to quantum mechanics. As is well known, quantum mechanics has led to a revolution in our thinking. In classical mechanics, "observables" are represented by numbers. The new point of view, taken by quantum mechanics, is to represent observables by operators. For example, the Hamiltonian H now becomes the Hamiltonian operator H_{op}. To this operator we associate eigenfunctions u_n and eigenvalues ε_n :

$$H|u_n> = \varepsilon_n|u_n>. \qquad (4.1)$$

The operator, H_{op}, acting on the eigenfunction, $|u_n>$, reproduces this function multiplied by the eigenvalue, ε_n. The eigenvalues correspond to the numerical values of the physical quantity associated to the operator, H_{op}. Once we have a complete set of eigenfunctions and eigenvalues, we have the "spectral" representation (we drop the subscript "op") associated to H:

$$H = \sum \varepsilon_n u_n><u_n. \tag{4.2}$$

Finding the spectral representation (or solving the eigenvalue problem) is the central problem of quantum mechanics; however, this problem has been solved in only a few situations, and most of the time we have to resort to perturbation techniques. We may start, as in the Poincaré theorem, with a Hamiltonian of the form

$$H = H_0 + \lambda V. \tag{4.3}$$

where we suppose that the eigenvalue problem can be solved for the "unpeturbed" Hamiltonian H_0. We look then for eigenstates and eigenvalues of H, which we could expand in powers of the coupling constant λ. It is here that contact with Poincaré classification can be made. For non-integrable Poincaré systems, the expansion of eigenfunctions and eigenvalues in powers of the coupling constant leads to the Poincaré catastrophe, due to the divergence associated to the small denominators. The relation between Poincaré's theorem and the quantum eigenvalue problem has been studied in a recent paper by T.T. Petrosky and the author [10].

Let us again consider the problem of quantum transitions (see Figure 6.)* When we try to solve this problem by conventional perturbation theory, we come to Poincaré's divergence associated in the example to denominators of the form

$$\frac{1}{\omega_1 - \omega_k} \tag{4.4}$$

---

* We in fact consider the simple version called the Friedrichs model in which virtual processes are omitted [15].

where ω_1 is the energy of the excited states and ω_k the energy of a mode of the radiation corresponding to wave vector k. To avoid the divergence, we have to give a meaning to the denominator in (4.4).

Here enters the basic element that makes Poincaré non-integrable systems "integrable" in a new extended sense. We introduce a "natural time ordering" of the dynamic states. To make clear what we mean, consider a trivial example. A stone can fall into a water pond and produce outgoing waves. We may also have the inverse situation in which incoming waves would eject a stone. (See Figure 8.)

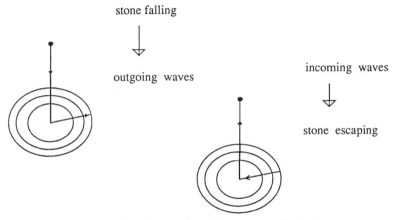

Figure 8. **Example of Temporal Ordering**

In fact, only one of the situations is realized: the natural time ordering is the falling stone first, the outgoing waves next. Similarly, to give a meaning to Poincaré's denominators, we have to time-order the dynamical states — the unstable atomic state first, the emission of radiation later. This corresponds to Bohr's picture in which the radiation emitted by the atom corresponds to a retarded wave.

More precisely, to the transition 1 —> k we associate the denominator

$$\frac{1}{\omega_1 - \omega_k - i\varepsilon} \qquad (4.5)$$

and to the transition k —> 1 the denominator

$$\frac{1}{\omega_1 - \omega_k + i\varepsilon}.$$ (4.6)

As we have shown [11,12, 13], this simple rule leads to the elimination of all Poincaré divergence when we integrate over the wavelengths of the radiation. It is a standard procedure in theoretical physics to express the difference between past and future through "analytic continuation." The general reader may just accept that we modify the Poincaré denominators differently according to the type of process to which they are associated. We obtain in this way complex solutions of Schrödinger's equation. The eigenvalues contain now an imaginary part corresponding to damping, and the eigenstates have a broken-time symmetry. We have chosen this simple example because in this case there exists a standard solution which is, however, not analytic in the coupling constant (the particle disappears from the spectrum [15]).

We can therefore compare our results with the standard treatment and see if our approach makes sense. We indeed recover all known results, but in addition, as we have states with broken time symmetry, we can introduce a functional which plays the role of Boltzmann's $\mathcal{H}$-function and which decreases monotonically when the particle emits the radiation and decays to the ground state. (See Figure 9.)

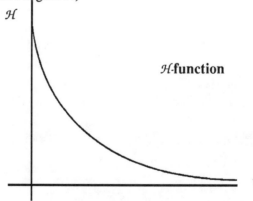

Figure 9. $\mathcal{H}$-function associated to the decay of unstable particle (see text)

We can also, at least as a thought experiment, perform a time-inversion at time t_0, after the start of the decay. The result is represented schematically on Figure 10. At the time of the inversion t_0, the $\mathcal{H}$-quantity has a jump ~ exp ($\gamma\, t_0$), where t_0 is the lifetime of the unstable state. Then $\mathcal{H}$ starts to decrease again: at $2t_0$ we have $\mathcal{H}(2t_0) = \mathcal{H}(t=0)$, and the decrease of $\mathcal{H}$

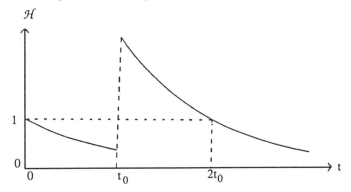

Figure 10. Time inversion experiment (see text)

continues until the particle has decayed. As we can associate an $\mathcal{H}$-function to the particle decay, the decay becomes an irreversible process

We may summarize what we have done as follows: to avoid Poincaré's catastrophe we have enlarged the type of transformations that lead from the eigenfunctions of the unperturbed Hamiltonian H_0 (see (4.3)) to the eigenvalues of the full Hamiltonian. In more technical terms, the situation is as follows: Poincaré considered only canonical (or unitary) transformations, which, among other properties, keep the eigenvalues of H real. We introduce more general transformations leading to complex eigenvalues. The specific choice of these transformations follows from our time ordering of the dynamical states.

The result is already of great interest in the perspective of the epistomological problems which plague quantum mechanics. Let us first recall that the basic equation of quantum mechanics is the Schrödinger equation for the wave function Ψ,

$$i \frac{\partial \Psi}{\partial t} = H_{op} \Psi. \qquad (4.7)$$

The equation, as is shown in all textbooks on quantum mechanics, is time reversible and deterministic. (We exclude some "pathological" cases related to weak interactions.) The physical interpretation of Ψ is that it represents a probability "amplitude." In contrast, the probability proper is given by

$$\Psi \bullet \Psi^{cc} = |\Psi|^2. \tag{4.8}$$

We shall come back in the next section to this transition from probability amplitudes, Ψ, to probabilities proper, $|\Psi|^2$.

Are there quantum jumps? This is quite a controversial problem. Schrödinger's equation (4.7) describes a smooth evolution. How then to include quantum jumps? In addition, (4.7) is time symmetric: if there is spontaneous emission, there should also be spontaneous absorption. The conventional Copenhagen interpretation is that quantum jumps result from our measurement. This would be rather strange, since all of chemistry and life are the result of quantum jumps. How could life be a result of our measurements? Our method solves this problem, as it associates to the quantum jump an irreversible event which can only occur in (our) future.

Quantum mechanics is probably the most successful theory of physics, and still, the discussions about its conceptual foundation have never ceased. I recommend a recent book by J.S. Bell, *Speakable and Unspeakable in Quantum Mechanics* [16].

Before we come back to these fascinating problems underlying the close connection between the conceptual foundations of quantum mechanics and dynamical instability, let us make the following remark. The example we have treated is a very simple one, as we could introduce a natural time ordering in the frame of the usual quantum description (the so-called "Hilbert space"). But in general, this is impossible (think about scattering where all states play a symmetrical role). We shall see in the next section that we have to introduce a natural time ordering on the level of the statistical description. This leads to the integration of LPS in quite general situations and to a new form of dynamics, which breaks radically with the past.

V. Poincaré's Theorem and a Statistical Formulation of Dynamics

From the example studied in Section IV, it should be clear how we may avoid Poincaré's catastrophe: It is through introducing into the theory a time ordering of dynamical states that leads to well-defined "regularization" procedures for the small denominators (see (4.5) - (4.6)). But how do we introduce this time ordering? Here, as we shall see now, we have to turn to the statistical description. Curiously our approach validates the way Boltzmann, more than a century ago, approached the kinetic theory of gases. But Boltzmann could not guess the emergence of the chaos theory and did not know that he was studying "non-integrable Poincaré systems." (He, as well as Maxwell, placed therefore his hopes in ergodic theory, which is indeed useful for the understanding of equilibrium, but not for dynamical purposes.)

In the early days of statistical mechanics, Gibbs introduced a quite fundamental concept, the "Gibbs ensembles." Instead of considering single dynamical systems, he considered a large number of dynamical systems evolving in the phase space associated to the coordinates $q_1...q_N$ and the momenta $p_1...p_N$ of the particles forming each dynamical system. (See Figure 11.)

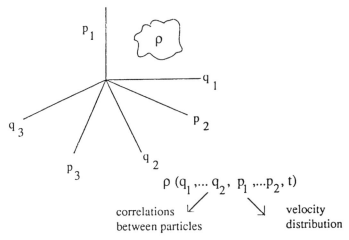

Figure 11. Gibbs ensemble

The description is then, in terms of the probability distribution ρ in phase space,

$$\rho(q_1...q_N, p_1...p_N, t). \qquad (5.1)$$

This description also remains meaningful for quantum systems. The probability distribution ρ is then called the "density matrix." Once we know ρ we can calculate both the velocity distribution of the particles as well as the correlations existing between the particles.

How then does time enter into this description?

Let us consider a classical gas. Particles collide and these collisions give rise to correlations. See Figure 12. First we have binary correlations, then ternary correlations, and, time going on, correlations involving more and more particles.

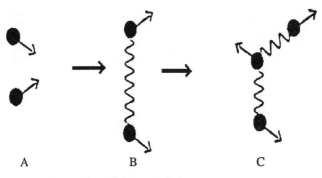

A: Binary correlations
B: Ternary correlations

Figure 12.   Flow of correlations

The formation of correlations is somewhat reminiscent of a couple which has a conversation (this would correspond to a collision). Even when the partners go away, the memory of their conversation remains. The information associated to this conversation is, time going on, spreading out to more and more participants.

Suppose we look at a glass of water. In this glass of water, there is an arrow of time that will in fact persist forever and corresponds to the creation of new correlations involving an ever-increasing number of particles. According to the correlations that exist between the molecules, we can distinguish "young" water from "old"! Computer experiments have been performed recently that show that binary correlations appear very rapidly. Ternary correlations involve longer time scales, and so on. This time-oriented flow of correlations breaks the symmetry involved in the classical description. Let us go from state A (of a many-body system) with no correlations at t=0 to a state B at time t involving multiple correlations. (See Figure 13).

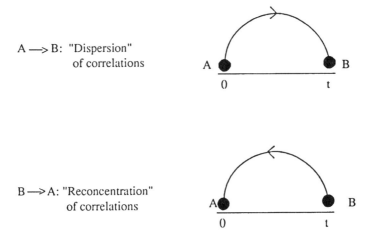

A —> B: "Dispersion" of correlations

B —> A: "Reconcentration" of correlations

Figure 13. Breaking of time symmetry (see text)

Gibbs's ensemble theory leads to an equation for the time evolution of the density matrix ρ

$$i \frac{\partial \rho}{\partial t} = L\rho \tag{5.2}$$

which is formally quite similar to the Schrödinger equation (4.7). L is the so-called Liouville operator, which can be expressed in terms of the Hamiltonian both in classical and quantum mechanics. As we mentioned,

$$H = H_0 + \lambda V. \tag{5.3}$$

This corresponds to a decomposition of the Liouville operator

$$L = L_0 + \lambda L_V. \tag{5.4}$$

To solve Liouville's equation (5.2) we need, as in the Schrödinger case, to solve the eigenvalue problem. (See 4.1)

$$L|f_n\rangle = l_n|f_n\rangle. \tag{5.5}$$

For integrable systems, there is no problem. Liouville's equation (5.2) is then of no special interest, as the problem reduces to the usual dynamical problems (finding trajectories or wave functions). However, the problem changes radically for non-integrable systems. Then the Liouville equation describes the emergence of chaos due to the destruction of the invariants of motion associated to the unperturbed system.

Again, Poincaré's theorem prevents us from finding solutions of (5.5) through unitary transformations (preserving the reality of l_n) which we could expand in powers of the coupling constant λ. As already mentioned, we solve this difficulty by introducing a supplementary element into the theory: the time ordering of the correlation. We then obtain a complex eigenvalue problem that can be solved and which leads to damping and to irreversibility through the occurrence of $\mathcal{H}$-functions (see Section IV). This new dynamics has some distinctive features which present a basic departure from the features of the dynamics of integrable systems as described in classical or quantum theory.

To understand in qualitative terms what happens, let us analyze more closely what is involved in the idea of "collisions." In fact, a collision corresponds already to a complex process in which particles come close, exchange energy through resonance, and depart. We can visualize a collision as a succession of states bound by resonance (see [8]). In a Hamiltonian system (the case of hard spheres is a limiting case which will not be considered here) a collision is not an instantaneous point-like event, but has an extension both in space and time.

As has been shown recently by T. Petrosky and the author, the spectrum of the Liouville operator L is essentially determined by the dynamics of the collisions. This implies a radical deviation from the usual methods of dynamics valid for integrable systems where the evolution can be resolved into a succession of instantaneous space-time events (remember Feynman diagrams).* For this reason the dynamics of LPS can only be formulated on the statistical level, as we cannot reduce it to trajectories, as in the classical case, nor to wave functions, as in the quantum case. This deviation from the great traditions of dynamics not so astonishing; we deal with an aspect of dynamics that is totally absent in integrable systems. It is, however, already present in the KAM theory, but there the behavior is so complex that it defies any quantitative description (we have to use qualitative criteria for the collapse of resonant tori as the result of the coalescence of resonances). It is precisely the main advance realized by the study of LPS to present a simple description of the physical processes due to resonances which lead to Poincaré's non-integrability.

Our approach has been confirmed by numerical calculations performed on simple examples of LPS. We may start with a statistical distribution (as close as we want to a point in phase space). We see then the system going through various stages corresponding to the appearance of Lyapounov instability (see (2.1)), folding in phase space and then diffusion due to "collisions."

The main point I want to emphasize again is that instabilities destroy the very notion of trajectory (or of wave function in quantum mechanics), as the basic description is now in terms of statistical ensembles.

---

*For the reader familiar with kinetic theory, let me mention that the traditional kinetic equations (the Fokker-Planck equations) contain <u>second</u> derivatives; this is precisely due to the description of the collision as a two-stage process.

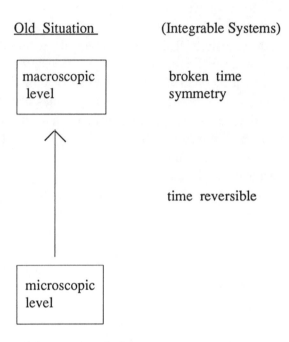

Figure 14. The Time Paradox (see text)

VI. Concluding Remarks

The integration of Poincaré's non-integrable dynamical systems leads for LPS to a new form of dynamics encompassing irreversibility (broken time symmetry) and exhibiting an increased role of probability, both in classical and quantum mechanics. The time paradox we have described in Section II is in this way eliminated. (See Figure 14.)

In the "old" situation, we had to bridge the microscopic time-reversible level to the macroscopic level equipped with an arrow of time (Figure 14). But how can time arise from no-time?

Now (Figure 15) we have a new microscopic level with broken time symmetry out of which, through averaging procedures, emerges the

New Situation

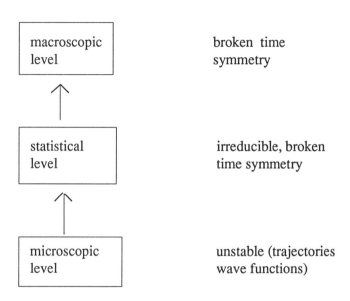

Figure 15. Elimination of the time paradox (see text)

macroscopic dissipative level. The "old" microscopic level has become unstable.

This leads to a better understanding of the role of chaos. In fact, there exist two quite different manifestations of chaos. When we study macroscopic equations that include dissipation, such as the reaction-diffusion equation or the Navier-Stokes equation for fluids, we are already facing situations for which the basic microscopic description belongs to LPS. In other words, the very existence of such equations presupposes "dynamic chaos." This is not astonishing; indeed, properties such as friction or diffusion involve exchange of energy through collisions. These macroscopic equations may lead to chaos (chemical chaos as turbulence). This dissipative chaos lies "on top" of the dynamic chaos. As we mentioned, dissipative chaos is part of self-organization as it appears in non-equilibrium and non-linear systems.

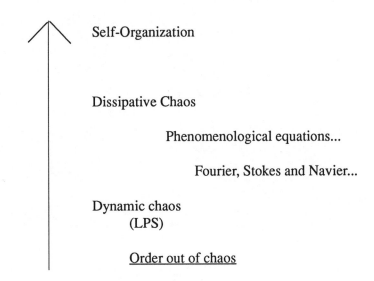

Figure 16.   Chaos and Dissipation

Examples of chemical coherence are oscillating chemical reactions. In short, therefore, macroscopic order as manifested in non-equilibrium is the outcome of dynamical chaos. Even the approach to equilibrium becomes the result of dynamical chaos. In all these cases, therefore, we have "Order out of Chaos" (see [1]).

Let us also mention that LPS are evolving systems. Once initial conditions are given, they go through various stages, such as those described by Lyapounov exponents, diffusional processes, etc. However, irreversibility is not related to Newtonian time (or its Einsteinian generalization) but to an "internal" time as expressed in terms of the <u>relations</u> between the various units which form the system (such as the correlation between the particles). We cannot stop the flow of correlations, as we cannot prevent the decay of unstable atomic states.

Nabokov has written: What is real cannot be controlled, what can be controlled is not real. This is also true here. In addition to solving the time paradox, the dynamical laws obtained through the integration of LPS lead to a number of consequences that go far beyond our initial motivation. We have already mentioned some relations with the epistomological problems of quantum mechanics in Section IV. We can now go further. As is well know, the basic quantity in quantum mechanics is the probability amplitude, which satisfies Schrödinger's equation (4.7), but we measure probabilities! Therefore, we need an additional mechanism to go from "potentialities" as described by the wave function to "actualities" as described by probabilities. In his introduction to *The New Physics*, Paul Davies [19] wrote, "At the rock bottom, quantum mechanics provides a highly successful procedure for predicting the results of observations on microsystems, but when we ask what actually happens when an observation takes place, we get nonsense! Attempts to break out of this paradox range from the bizarre, such as the many universes interpretation of Hugh Everett, to the mystical ideas of John von Neumann and Eugene Wigner, who invoke the observer's consciousness. After half a century of argument, the quantum observation debate remains as lively as ever. The problems of the physics of the very small and the very large are formidable, but it may be that this frontier — the interface of mind and matter — will turn out to be the most challenging legacy of the New Physics." It is interesting that the solution to this fundamental problem may come from dynamical instability and chaos, as in our new dynamical description we deal directly with probabilities. In this case, the breaking down of the superposition principle of quantum mechanics in LPS is due to dynamic instability, without any appeal to esoteric considerations, such as the many-world theory or the existence of the new universal constant leading to a collapse of the wave function for macroscopic systems. We come to a realistic formulation of quantum mechanics eliminating the appeal to any observer situated outside physics.

This century has been dominated by two new conceptual frameworks: quantum mechanics and relativity. As has often been emphasized (see, e.g., M. Sachs [20]), the intrusion of subjectivistic elements through the measurement process leads to difficulties when we want to combine quantum theory and relativity. However, non-integrable dynamical systems are likely also to alter relativity, as the basic dynamical events (the collisions) no longer correspond to instantaneous and localized space-time events.

Therefore, I believe that we are indeed at the beginning of a "New Physics." Until now, our view of nature was dominated by the theory of integrable systems, both in classical and quantum mechanics. This corresponds to an undue simplification. The world around us involves instabilities and chaos, and this requires a drastic revision of some of the basic concepts of physics.

Let me conclude by expressing my conviction that, in the future, the non-integrability theorem of Poincaré will be considered as a turning point somewhat similar to the discovery that classical mechanics led to divergences when applied to the black-body radiation. These divergences had to be cured by quantum theory. Similarly, Poincaré's divergences have to be cured by a new formulation of dynamics in the sense I have tried to describe in a qualitative way in this presentation.

Appendix

Since this lecture was presented, great progress has been realized. Complex spectral theory is now firmly grounded in "rigged Hilbert space" formulation. As a special case, this leads to a formulation of the spectral theory of chaotic maps. The laws of chaos correspond to <u>irreducible</u> statistical formulations, as they correspond to the evolution of probability distribution (and are not applicable to individual trajectories). This leads to a new definition of chaos as characterized by irreducible representations, which can be extended to Hamiltonian (classical or quantum) systems. In this way, we obtain for chaotic systems results linking directly dynamics, statistical mechanics and thermodynamics, which would be impossible to derive from trajectory theory (Newton) or traditional quantum theory (Schrödinger). Most of these results can be found in the papers listed in the references.

Acknowledgements

This work is the outcome of many years of work of the Brussels-Austin group. It is impossible to quote all those who have contributed, but my special gratitude goes to Cl. George, F. Mayne, and T. Petrosky.

We want also to thank The University of Texas at Austin, The Solway Institutes, the Department of Energy (DOE grant DE-FG05-88ER13897), the Welch Foundation (grant WELCH FDN F-365) and the European Communities Commission (Contract No. PSS*0143/B).

References

[1] I. Prigogine and I. Stengers, *Order out of Chaos*, Bantam Books, New York, 1984.

[2] This question is discussed in Ilya Prigogine and Isabelle Stengers, *Entre le Tempset Eternité*, Fayard, Paris, 1988.

[3] See the excellent introduction by M. Tabor, *Chaos and Integrabil ity in Nonlinear Dynamics*, J. Wiley, New York, 1989.

[4] G. Nicolis and I. Prigogine, *Exploring Complexity*, W.H. Free man, N.Y. 1989.

[5] P. Coveney and R. Highfield, *The Arrow of Time, Allen,* London, 1990.

[6] For a summary, see Ref. [4].

[7] H. Poincaré, *Methodes Nouvelles de la Mecanique Céleste*, Gauthier Villars, Paris, 1892, reprinted by Dover, New York, 1957.

[8] For a simple presentation, see Ilya Prigogine, *Non Equilibrium Statistical Mechanics*, Wiley Interscience, New York, 1962. in Foundations of Phyics, 1990.

[9] Let us only mention a few references in addition to [8]:
   [a] I. Prigogine, *From Being to Becoming*, W.H. Freeman, New York, 1980.
   [b] R. Balescu, *Equilibrium and Non Equilibrium Statistical Mechanics,*Wiley, New York, 1985.
   [c] Cl. George, F. Mayné, and I. Prigogine, *Adv. in Chemical Physics*, 61(223), 1985.

[10] T. Petrosky and I. Prigogine, *Physica* 147A (439), 1988.

[11] H. Hasegawa, T. Petrosky, I. Prigogine, and S. Tasaki, "Quantum Mechanics and the Direction of Time," *Foundations of Physics,* 21, No. 3 (March 1991).

[12] T. Petrosky, I. Prigogine, and S. Tasaki, "Quantum Zeno Effect," *Physica A* 170, (1991) 306-325.

[13] I. Prigogine, T. Petrosky, and S. Tasaki, to be published.

[14] T. Petrosky and H. Hasegawa, "Subdynamics and Nonintegrable Systems," *Physica A* 160 (1989) 351.

[15] See, e.g., G. Balton, *Advanced Field Theory,* Wiley Interscience, New York, 1963.

[16] J.S. Bell, *Speakable and Unspeakable in Quantum Mechanics*, Cambridge University Press, 1987.

[17] The idea of "dynamics of correlations" was introduced in Ref. [8].

[18] H. Hasegawa and B. Saphir, *Physics Letters A* 161, (1992) 471.
P. Gaspard, *J. Phys* A 25 L483 (1991).
I. Antoniou, S. Tasaki, *J. Phys.* A, in press.

[19] *The New Physics,* edited by Paul Davies, Cambridge University Press, 1989.

[20] M. Sachs, *Einstein versus Bohr,* Open Court, Illinois, 1988.

[21] T. Petrosky and I. Progogine, *Physica* 175A, p. 146 (1991), PNAS to appear 1993.

DISCUSSION

QUESTION FROM AUDIENCE: With regard to your theory of irreversibility of time, what's the connection with the big bang theory?

PRIGOGINE: First of all, let me make a general remark. In many books, even those written by very distinguished writers, when they speak about the cosmic direction of time, they also speak about a biological direction of time, a cultural direction of time, and so on. What we have tried to do is to show that all these "hours" of time, all these symmetry-broken times, have the same dynamical origin and correspond to the same type of fundamental description. Now when we come to the origin of the universe, then the question becomes more phenomenological because we know much less about the origin of the universe than we know about simple dynamical systems. That is, by the way, the reason why I do not have great sympathy for explaining irreversibility by cosmological evolution. I think that cosmological evolution is just one example.

But now, let's come back to cosmological evolution. Here the question is: What is the meaning of the big bang, and what is the likely nature of future evolution? Well, here this would require another lecture which I hope to do sometime, but not today. It prerequires the question: What is the mechanism of the big bang? Is the big bang a singularity or an instability? My personal feeling is that it's an instability not a singularity. In other words, the early universe was a highly fluctuating quantum vacuum in which you could form, let's say for example, many black holes, which acted as membranes and then produced matter as we know it. If this is so, then the very process of the creation of the universe is an irreversible process. And it is this irreversibility you find in the dual structure of the universe. On one side the universe is full of this order. It is full of the residual black-body photons which are floating around, as you know, one elementary particle for each ten to the exponent eight or nine photons. On the other hand, elementary particles like hadrons have an unbelievably complex structure. I would call them the proteins of elementary particle physics. So you have this superposition of the reversible and structure-forming processes which are the price of formation of the universe is the price of entropy. To give you a very, very simple idea which I like to explain in a somewhat childish way, I would say that at the moment of the creation of the universe, it becomes small pieces, as when I tear up this paper. Of course, this creates a lot of new degrees of freedom. These new degrees of freedom create the new entropy of the universe and, therefore, the creation of the universe is the primary example of an irreversible process. Of course chaotic systems provide another illustration.

QUESTION FROM THE AUDIENCE: I understand that chaos has descriptive power. Does it or will it have predictive capabilities?

PRIGOGINE: In my opinion, it has already a very important predictive power. Once you understand when chaos arises, you can also understand how you can change the condition for the appearance of chaos, how you could make a fluid stable in conditions where it should become turbulent, how you could retard the appearance of chaos in chemical reactions. This is why I don't like so much the word "chaos," which is so popular. It emphasizes the negative aspects. In fact, "chaos" gives us a handle on an enormous amount of phenomena which we could not even imagine how to handle before. Essentially everybody knows that we can predict the position of the earth for long periods of time, for five million years. And everybody knows that we cannot predict the weather for long periods of time, but at best for ten days to three weeks. That's related to the characteristics of instability of the weather. The weather, like the climate, is a two-stage phenomenon: good weather/bad weather, cold climate/warm climate. Once we begin to understand these transitions, we can hope to work out better conditions of life for men in the next century. Our earth is in a very poor condition not only because we are destroying it to some extent, but also because geological conditions have been deteriorating since the last glacial period. One of the main aims of science is, of course, to make the world more habitable, more acceptable for men. We have to go beyond conservation. We have to go beyond anti-pollution, and in going beyond, we have to understand the laws which are fundamental for the biosphere and climate and so on. These laws involve chaotic systems. Therefore, the science of chaos is our great hope for improving our environment, for improving the conditions of life of man in nature.

QUESTION FROM THE AUDIENCE: What relation, if any, is there between dissipative structures and fractal geometry?

PRIGOGINE: I would say that while fractal geometry emphasizes the geometrical aspect, dissipative structures emphasize the non-equilibrium aspect. Both aspects, of course, coexist in some examples. There are many problems in which diffusion in chemical reactions give rise to fractals. That is precisely one of the beautiful outcomes of the research by Benoit Mandelbrot. Therefore, I think they are simply emphasizing complementary aspects. Also, if you speak about attractors in dissipative chaos, then, as was already mentioned, attractors in dissipative chaos are generally not points, not lines, but are more complex, and their complex geometrical nature is expressed by the fact that they are fractals.

I would like to make a general comment here. In the neolithic age—and I have always been very interested in neolithic art, in neolithic culture—people were interested in simple forms. In other words, you see already in the paleolithic age all these beautiful spirals and so on. In the first gifts which you find among the Neanderthal, fifty to seventy thousand years ago, are very spherical small stones. In other words, beauty was associated with geometry, with simple geometry, and this is in fact the precursor of what became geometry later. Today, in a sense, we can go beyond this idealization, and we can have a look on a richer nature. We are not obliged to speak about spheres. We can speak about much more complicated, self-similar objects. I still think we have to keep our admiration and our veneration for the simple geometrical objects.

QUESTION FROM THE AUDIENCE: I have one last question, a disarmingly simple little question to Dr. Prigogine. What is time?

PRIGOGINE: Certainly time is not motion. Time is a global property. We have to deal with two aspects of time. First of all, there is a Newtonian, Einsteinian time. It is the time of the Schrödinger equation. And there is nothing wrong with that. This describes perfectly well a class of situations associated to integrable systems. This is essentially motion. Now motion is timeless because, in a sense, as I said in the beginning, if you have a pendulum going in this direction or that, there is no time ordering. This watch is not time ordered. I'm time ordered, but the watch is not time ordered. Therefore, the question is, how time order enters into the basis of physics. It enters through topological relations. It enters by the fact that the units become correlated, that they are no more independent, that they have some properties in the global structure. It is not something which you can analyze on the level of single particles. It is something you can analyze on the level of the relations. And this, I think, as I mentioned, is very similar to human relations. After all, the time that is passing is to some extent related to our occupations. Occupations are relational things—human relations, friends, loves, all that makes life. And this is really what gives irreversibility. Therefore, irreversibility is something that goes beyond what classical mechanics had introduced.

The curious thing is that the classical mechanics of interglobal systems is complete. You cannot add to it. If you did, you would destroy it. For a long time, people tried to take it out from classical mechanics. But you cannot, because it's complete. The interesting fact is that since Poincaré and Kolmogoroff, we have dynamical structures which are incomplete, and we can now add a new element. There is incompleteness manifested in a very nasty way. It is manifested in nonintegrability, in divergence. But, the divergence is in our mind. It's not in nature. When you make an experiment on a computer, the experiment will not show you infinities, but what the experiment shows you is precisely irreversibility and stochasticity. Therefore, nature solves the problem of nonintegrability. Nature has broken time symmetry and is stochastic. Only our equations are too narrow to include it. I think if chaos is really becoming a new science, it is because of its role in opening dynamics to include the direction of time.

WHAT IS CHAOS?
STEVE SMALE

In an attempt to explain my own part in chaos theory, let me start with a few words on the role of mathematics in science.

There are at least three (overlapping) ways that mathematics may contribute to science. The first, and perhaps the most important, is this: Since the mathematical universe of the mathematician is much larger than that of the physicist, mathematicians are able to go beyond existing frameworks and see geometrical or analytic structures unavailable to the physicist. Instead of using the particular equations used previously to describe reality, a mathematician has at his disposal an unused world of differential equations, to be studied with no *a priori* constraints. New scientific phenomena, new discoveries, may thus be generated. Understanding of the present knowledge may be deepened via the corresponding deductions.

Those who anticipated general relativity illustrate my point. It was the mathematicians who first saw that geometry did not have to be Euclidean, that there could exist, in principle, different possible worlds. These discoveries were crucial for Einstein's great achievements.

Another example of this kind of contribution is found in the history of chaos. Physicists (and many mathematicians as well) were slow to recognize chaotic phenomena because they were oriented towards solving particular equations and analyzing particular solutions. But even 100 years ago, Poincaré had seen the possibility of homoclinic points by transcending that methodology. (My feeling is that homoclinic points lie at the heart of chaos; later I will describe them and defend that point of view.)

A more recent development of the theory of homoclinic points occurred when mathematicians took the bolder step of looking at the completely arbitrary differential equation, no longer tied to the physical world. This gave the freedom to create geometric constructions, unhampered by specifics. In this broader picture, the centrality of homoclinic phenomena became clear and an appropriate analysis was carried out. That analysis was then applied to the traditional equations of engineering systems.

The second way in which mathematicians contribute to science has to do with the consolidation of new physical ideas. One may express this as the proof of consistency of physical theories. Examples here include the mathematical foundations of quantum mechanics with Hilbert space, its operator theory, and corresponding differential equations.

Closer to the subject of our talk is the proof of the Feigenbaum conjectures by Lanford and the development of the geometric Lorenz attractor by Williams, Guckenheimer, and Yorke.

The third way in which mathematicians contribute to science is by describing reality in mathematical terms, or by simply constructing a mathematical model. I feel that this purely phenomenological approach has less importance for science than the first two ways I have described. My criticism of catastrophe theory* was along these lines; and now I feel some chaos and fractal practitioners are turned too much in this direction.

Let us turn to the question, "What is chaos?" Certainly the typical answer, "sensitive dependence on initial conditions," is reasonable. But to be relevant to physics, more is required: the property should be robust or preserved under perturbations of the system. A response that involves a mechanism and meets the above criteria is:

Chaos is the presence of a (transversal) homoclinic point.

(One would allow some borderline case where this might not be exactly right.)

As a step towards understanding this statement, the mathematical meaning of homoclinic point needs to be given. A homoclinic point is a state which tends to a certain equilibrium in both forward and backward time under the dynamics. Let us be more precise. All of the following considerations apply to n-dimensional dynamics, but for simplicity of exposition we take states to be in a plane.

Consider dynamics with time discrete and reversible. Thus time t takes values 0 (now), 1 (1 unit forward), 2,3,... and for the past $t = -1,-2,...$. For a state z in the plane, z_t denotes the state at time t which was initially z; so $z = z_0$. Then $z_{t+1} = T(z_t)$ where T is the transformation of the plane onto itself giving the dynamics.

---

* *Bulletin of the American Mathematical Society* 84 (1978), pp. 1360 - 1368. Reprinted in Steve Smale, *The Mathematics of Time,* Springer - Verlag, New York, 1980 —Ed.

WHAT IS CHAOS?

As an example let $T(x,y) = (2x,(1/2)y)$, where (x,y) are the coordinates of a point z in the plane. The dynamical situation can be represented by the diagram

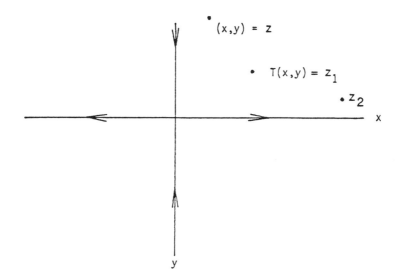

Note especially how, under time, states are contracted in the vertical direction toward the x-axis and simultaneously expanded away from the vertical axis. Any state on the y-axis remains there for all time and the same is true for the x-axis. This property may be stated as "the x-axis (to be called W^u) and y-axis (to be called W^s) are *invariant* under the dynamics." A state on the y-axis tends to the origin as time gets larger and larger.

In this example the dynamics, or T, is linear, and no chaos can be present. Yet understanding this simple example is a good prelude to understanding the deeper examples to come.

Returning to the general case, we define a state p in the plane to be an *equilibrium* of a dynamics T provided that $T(p) = p$. Thus, p is stationary in time. In our linear example $p = (0,0)$ is the unique equilibrium. A crucial development in the recent theory of dynamical systems is the emphasis on the set of all states approaching an equilibrium p as time increases indefinitely (i.e., p such that $z_t \to p$ as $t \to \infty$). Call this set the stable manifold W^s of p. The linear example above has the vertical axis as its stable manifold.

Similarly one defines the unstable manifold W^u of p as the set of points which tend to p as time goes backward indefinitely ($z_t \to p$ as $t \to -\infty$). In our example, W^u is the horizontal axis. From the definitions it follows that both W^s and W^u are invariant under the dynamics.

A point q different from p which belongs to both W^s and W^u is a *homoclinic point*. Thus q will tend to p in both forward and backward time. In our example it is clear that there are no homoclinic points. This is an illustration of the relation between chaos and homoclinic points (lack of both!).

Now imagine that the linear dynamics T is modified a bit to a nonlinear transformation of the plane onto itself. Our diagram might change to:

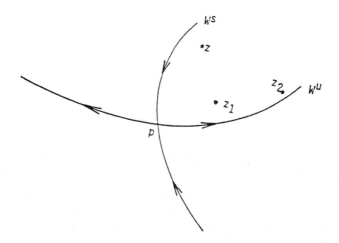

The stable manifold W^s of the fixed point p is no longer straight, but still remains a curve nicely imbedded in the plane. The same applies to W^u. There are no homoclinic points. The next step is to imagine a larger modification of T, so large in fact that W^s and W^u cross as in the following diagram.

WHAT IS CHAOS?

This crossing point q is a homoclinic point. A technical condition, necessary for the subsequent analysis, is that q is a *transversal* crossing point: W^s and W^u are not tangent at q. Let us explore some consequences that follow merely from the existence of this transversal homoclinic point q.

By the invariance of W^s and W^u, q_1 lies on W^s and W^u, where q_1 is q at one unit of time later.

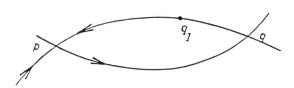

This forces the picture:

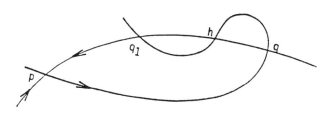

Note that another homoclinic point h must be present. Similarly going backwards in time we see:

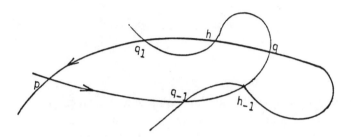

Eventually in time, the curve on W^u between h and q, becomes stretched along W^u, as between h_1 and q_2 in the next diagram.

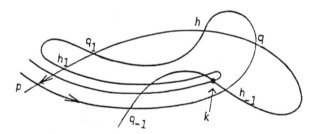

This part of W^u is forced to meet up with W^s, the part of W^s between h_{-1} and q_{-1} to create an additional homoclinic point k. The above analysis can be applied to k.

One can begin to understand Poincaré's statement in 1899, referring to homoclinic points [5, p. 839]: "On sera frappé de la complexité de cette figure, que je ne cherche même pas à tracer. Rien n'est plus propre à nous donner une idée de la complication du problème des trois corps et en général de tous les problèmes de Dynamique..." **

Now that we have seen some of the complications that the existence of a homoclinic point forces upon us, we turn to the question, "Is there some way of making an analysis of the dynamics in this environment?" Toward that end consider the following curved rectangle containing part of W^s.

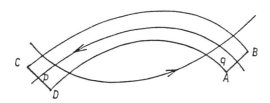

The dynamics contracts along W^s and stretches along W^u, and also this contracting and stretching applies to points in the neighborhoods of W^s and W^u, respectively. Thus at a certain time later the curved rectangle ABCD has become the new curved rectangle A B C D in the next diagram.

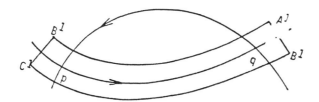

**One will be struck by the complexity of this figure, which I am not even going to attempt to draw. Nothing can better give us an idea of the difficulty of the three-body problem, and of dynamics problems in general..."—Ed.

Next we straighten out W^s to get the following diagram containing the dynamics of the homoclinic point.

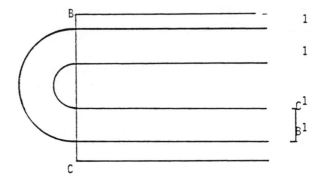

This dynamics is called the *horseshoe*, and we have argued that a homoclinic point implies the horseshoe dynamics. The converse can also be proved.

Note also we might have given a direct geometric construction of the horseshoe, not assuming the existence of a homoclinic point.

The horseshoe is quite amenable to a detailed analysis. We will indicate quite briefly how that can be done.

Consider the set of states which at one unit of time later lie on the intersection of the horseshoe and the square. This is shaded in the next diagram.

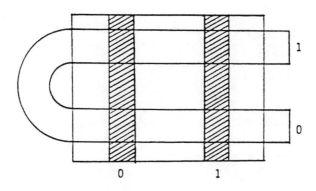

The two vertical strips we have labeled 0 and 1. If a state z has the property that z_t is in the square for all $t = 0, \pm 1, \pm 2,...$ then we assign to it a sequence of 0's and 1's with "a marker," e.g., ... 0111.011100 ... = ... $S_{-1}.S_0S_1S_2$ ... by

$$S_i = 0 \text{ if } z_i \in K_0.$$
$$S_i = 1 \text{ if } z_i \in K_1.$$

Here K_O is the horizontal strip labeled 0 and K_1 the horizontal strip labeled 1 in the diagram. It may be shown that this assignment is a 1-1 correspon
dence between the states and the sequence of symbols. The corresponding dynamics on the sequences moves the period forward one place. This allows the original dynamics of the homoclinic situation to be studied via the combinatorial problem of moving the period in the space of sequences of zeros and ones.

This construction, the horseshoe, has some consequences. First, it yields the fact that homoclinic points do exist and gives a direct construction of them. Second, one obtains such a useful analysis of a general transversal homoclinic point that many properties follow, including sensitive dependence on initial conditions — "a large number," anyway. Third, one can prove robustness of the horseshoe in a strong global sense (called structural stability).

Summarizing, we have indicated why homoclinic points imply chaos. The converse is less formalizable, but I believe it is basically true. However, there is a stronger notion of chaos, which we might call full chaos, where the set of initial conditions with sensitive dependence has positive or even full measure. In this case the corresponding homoclinic point could well be associated with a chaotic attractor, such as the Lorenz attractor. We stop short of pursuing the discussion in that direction here.

Let me end these notes by making some remarks about further reading. (1) A history of the horseshoe may be found in [4] and [7]. The mathematics of the horseshoe is found in [3], [6], [7]. (2) Some of the main outstanding problems in chaos can be found in [8]. (3) My own work in recent years has been in the areas of complexity and computation. But this is related to chaos through such questions as the decidability of the Mandelbrot set. See [1] and [2]. (4) My favorite book that gives an introduction to chaos is [3].

References

1. Blum, L.; Shub, M. and Smale, S., "On a theory of computation and complexity over the real numbers: NP-completeness, recursive functions and universal machines," *Bulletin of the American Mathematical Society* 21(1989), 1-46.

2. Blum, L. and Smale, S., "The Gödel incompleteness theorem and decidability over a ring," Preprint, Berkeley (1990).

3. Devaney, R., *An Introduction to Chaotic Dynamical Systems*, Addison-Wesley, Redwood City, California, 1989.

4. Gleick, J., *Chaos*, Viking, New York, 1987.

5. Poincaré, H., *Les Méthodes Nouvelles de la Mècanique Céleste*, Vol I-III, Gauthier-Villars, Paris, 1899.

6. Smale, S., "Diffeomorphisms with many periodic points," in *Differential and Combinatorial Topology* (ed. S. Cairns), Princeton Univ. Press, Princeton, New Jersey (1965).

7. _____, *The Mathematics of Time*, Springer, New York, 1980.

8. _____, "Dynamics retrospective: great problems, attempts that failed," for *Non-linear Science: The Next Decade*, Los Alamos, May 1990.

DISCUSSION

GLEICK: I wonder whether I was the only person who was disappointed when Steve Smale stopped short of offering his opinion on whether he feels that current developments in chaos and fractals have been too much descriptive and phenomenological. In case I wasn't, maybe you could take the opportunity to say.

SMALE: It's a little complicated to do that in a short time. The simple answer is, some are and some aren't. It may be more fruitful to talk about those developments that are more enduring. I could talk on that for a while, but otherwise, I'm not sure what to say. Certainly, I do see a lot of literature on chaos and fractals I would regard as kind of superficial. Not a lot, but a certain amount, I would say is pretty superficial. I see a lot of beautiful things happening, too. I see both.

PRIGOGINE: Professor Smale is, I understand very much your emphasis on homoclinicity, but, why reduce chaos to homoclinicity? There may be other mechanisms like the one which I discussed this morning, resonances, which are all mechanisms for chaotic behavior as seen on the computers and as confirmed in the sense of sensitivity to initial conditions.

SMALE: All I can say is from my experience in this kind of dynamics I do not see resonances as a key thing to chaos. Resonances certainly play a big role, especially in the Hamiltonian formulation, which I think is not the place to look first of all for chaos. It's more in the dissipative or dissipative with forcing kind of a dynamics. In that kind of dynamics, at least from experience and examples, I would say resonances are not so clearly connected with chaos. That's my opinion.

PRIGOGINE: Mark my disagreement. In other words, I believe that it's very important to go from a Newtonian dynamics to dissipative chaos, and in a Newtonian dynamics it's a matter of fact that you find chaos or sensitivity to initial conditions is related to resonances and then to overlapping resonances. There's a lot of things to be done. I would not brush aside the problem of resonance.

SMALE: Let me respond once more. I think in Hamiltonian mechanics, Hamiltonian dynamics, one does, of course, have lots and lots of resonance. But also one has typically lots of homoclinic points. And to me it's the homoclinic points that are related to the sensitive dependence on initial conditions as I would define it, exponential increase of error.

MANDELBROT: In your introduction on the role of mathematics, I was amused to hear Poincaré quoted as a mathematician. After all, he was a professor of theoretical physics and a journalist who just dabbled in mathematics. The second person you mentioned, who provided the foundations of quantum mechanics, was also a very peculiar mathematician since he was, in a certain sense, hounded out of the mathematics profession. Therefore, I think what you did was to describe the influence on chaos of two very great men of the past, and those two men should not be appropriated by mathematics.

QUESTION FROM THE AUDIENCE: We know from your work on the Poincaré conjecture how dependent topology is on dimension 5 or higher. Does chaos theory display a similar dependence on dimension?

SMALE: I would say primarily, no. There are some theorems in dynamical systems, some theorems related to chaos, where topology plays an especially big role. In those theorems Poincaré's conjecture for dimension 5 does play a role, but I would say primarily for the dynamical phenomena, no.

QUESTION FROM THE AUDIENCE: On mathematics and physics do you believe that all of physics is determined in some yet unseen level by mathematical logic, i.e., does mathematics describe physics, or does physics describe the inevitable result of mathematics?

SMALE: Well, I don't have a lot of answers to that question, but I would surely say that it's not true that physics can be reduced to mathematics. That's for sure. Or, in any kind of a version of that statement, I would say no.

WHAT IS CHAOS?

PRIGOGINE: I think you said something very deep and very beautiful at the beginning of your lecture when you said that the world of mathematics is in a sense a less constrained world. In other words, there are many possible worlds and the role of the physicist is in a sense to make choices between all the possibilities which are intellectually possible. For example, you can perfectly imagine a world formed only by integrable systems without chaos. It's a perfectly coherent world. You can imagine a world formed by atoms which are randomly colliding, and that's all. But our world is not an arbitrary world, and therefore creation is, in a sense, the role of the physicist, who constrains the various possibilities opened by mathematics and from time to time tries to use some mathematical notions which had never been used before.

FEIGENBAUM: I'd like to make a comment loosely related to that. One of the points that Professor Smale made at the beginning of his talk was the ranking as #3 of the role of description, and in some way that's not necessarily the circumstance, in my mind, in physics. If one thinks of the development of mechanics, there are two fundamentally distinct things. One is called kinematics, and the other is called dynamics. The invention of kinematics by Galileo is by far the most profound part of the subject, and one of the ways that a physicist often thinks is that it's important to figure out the right setting in which you describe something and then how you should describe it. It's a profound statement to say that an object is a set of pieces, each of those, loosely speaking, made out of a point which lives in a certain 3-dimensional continuum and propagates according to a continuous time. That's a profound statement of description, and the realization that that sort of description was what was at stake in describing things is one of the profound accomplishments in erecting the structure of physics. So in some ways description could be arbitrarily important in the development of the science.

SMALE: I said these were overlapping characterizations of the ways in which mathematicians can contribute. A description can also carry great insights, a description can also be a discovery at the same time. By saying "purely descriptive," I meant to exclude that kind of insight. Benoit Mandelbrot, at lunch yesterday, gave me a very good example of something that was descriptive: Kepler's use of ellipses for planetary orbits. That was descriptive, but also it was an elegant, simple description. It was a description and a great discovery at the same time, a great insight. So I do say that those ways are overlapping.

GLEICK: I'd like to ask Professor Smale whether he thinks it's worthwhile making a distinction, within the concept of proof, between proofs that deliver some insight and proofs that don't? That is, one would hope that a proof that Lorenz's equations actually contain the geometric version of the Lorenz attractor would, as you suggested, offer some insight as well as simply confirming the suspicions of a thousand physicists. On the other hand, when one thinks of the proof of, for example, the four-color map theorem, a proof that involves thousands of pages of computer analysis and can't be comprehended by any individual, I think one would concede that that's a proof that doesn't provide any particular insight. Is that a useful distinction, and is it a distinction that mathematicians recognize? Is one sort of proof considered more desirable? Are there any criteria by which one can sort out insight-providing proofs from less useful proofs?

SMALE: Oh, I don't know about an answer to the last question about criteria for distinguishing proofs, at least any kind of formal criteria. Of course, I agree with you that an answer to the question I asked about the Lorenz attractor is likely to give us something important. I think it's likely just in view of the context of that problem. It's likely to give us some kind of mechanism for establishing chaos in more general systems. The problem has put up resistance and it's been around for so long that there must be some kind of stumbling block. In dealing with that, confronting that, we're likely to be led to some kind of more general conceptual criteria for saying that this or that physical system will be chaotic. And I think that could be very, very important for science. It's not sure. It could be a very superficial proof without insight. It's possible, but I think not so likely. And then there are degrees of insight, too, so it's not too simple a question.

PEITGEN: I think that James Gleick has touched a very important point about mathematics. I think that proof, in the first place, does not desire to find out whether something is right or wrong. That's not the role of mathematics. I think the role of proof is to find out why something is right or wrong, and in that sense, of course, proof means to deliver insight into something. In that sense, of course, proof means much more than just the mechanical decision by a computer whether a certain number is of that kind or another kind. I think that it's essentially the soul and role of mathematics to somehow bring order into the complexity of the universe of models which we make all the time. When we do science, we always make a model of what we see. When we are in front of a typewriter, at first we don't understand how the typewriter works, but soon we find out how it works because we make an internal model in our brain. And there are so many models floating around and that we need *order* to somehow compare the models and not get lost, and I think that's the primary role of mathematics and of proof.

SMALE: Fine. Makes sense.

MANDELBROT: When a problem has remained open for a very long time, it does not at all follow that the proof brings insight. There's an example which I like very much, the problem of minimal surfaces due to Plateau. A friend of mine proved Plateau's conjecture. A hundred and fifty years after the fact, she got great acclaim for it because it proved to be very difficult, but she says on every occasion her belief in the result didn't change in any fashion. And she learned nothing from her proof. But she learned a great deal, in contrast, for making computer experiments on minimal surfaces, and that's one part which you didn't discuss. You spoke of the influence of Poincaré and von Neumann on physical problems. And again, I don't count them as mathematicians, either of them. But you didn't speak of the influence of physics on mathematics, and I think the two are very strongly linked. In the case of minimal surfaces, the problems were not reopened because somebody solved a very long-standing open question, but because somebody had the good sense of using the computer very intelligently and discovering minimal surfaces and new conjectures about them. The field revived because of this second act. The same person did both, so there's no issue of personalities, but the fruitful thing was the second.

SMALE: First of all, I'll say that there are many, many, many important things I didn't say in my talk. So lots of these questions, like the influence of physics on mathematics, I didn't try to deal with. On the other hand, I certainly wouldn't make any absolute statements about the existence of a problem for a long time making its solution more important. But I would say there's a tendency in that direction. One of the first persons to get the Field's medal, Douglas, won it for his solution of Plateau's problem. In fact, his work has been very influential and conceptual.

PRIGOGINE: The problem with proof is, of course, such an enormous problem that it is difficult to make any comments, because we know that there are problems which are undecidable. Therefore, essentially so far as I can see, the problem of proof is to relate a specific property to a more general sense of those things which you admit. Therefore, when you cannot prove something in the frame of these general things, then the question is, can you generalize? Can you change the initial conditions from which you have started? I tried to give you an example in my lecture. Some systems are nonintegrable, and you can prove that they are not integrable—Poincaré has proved it. But you can then ask, can you not change your perspective to make them integrable in a different sense. And it's the same story about roots of algebraic equations and so on. Therefore, the question of proof is really of great importance in enlarging the intellectual horizon in which you are working.

QUESTION FROM THE AUDIENCE: Just how does one turn a sphere inside out without cutting or folding it?

SMALE: That's one question I will not try to answer at all. There's a movie on turning a sphere inside out, which is fairly well known. It gives the answer.

CHAOS AND COSMOS:
A THEOLOGICAL APPROACH
John Polkinghorne

The earth was without form and void and darkness was upon the face of the deep; and the Spirit of God was moving on the face of the waters.   Gen. 1, 2.

It is a commonplace that the dynamical theory of chaos proves to have applications in many branches of science. It might seem more surprising that it would be of relevance to theology. The reason that this is so lies in the twin roles of theology. One role is concerned with intellectual reflection upon the data provided by the religious experiences of humankind. Here theology stands alongside many other forms of rational inquiry into the way things are: science's investigation of the physical world, ethics' investigation of our moral experience, aesthetics' investigation of our experience of beauty. Each of these inquiries has its own domain of data and its own consequent autonomy: each has a cousinly relation to the others as they seek a rationally-motivated understanding of what is going on. All are required if we are to do justice to the richly-layered world in which we live.[1] Theology operating in this mode is what is often called systematic theology. It can neither tell science what to think (as the creationists wrongly suppose) nor be told by science what it should think (which would be the error we usually call scientism).

Valuable and indispensable as these separate rational inquiries undoubtedly are, they are not enough to satisfy our thirst for understanding. They are looking at different aspects of the world of our experience, but that world is surely one and we shall not be fully content till we are capable of integrating our views of it. I am a passionate believer in the ultimate unity of knowledge and so I am committed to the quest for that single view. The

---

[1] J. C. Polkinghorne: *One World* (SPCK, 1986; Princeton University Press, 1987); *Reason and Reality* (SPCK, 1991), ch. 1; J. R. Carnes: *Axiomatics and Dogmatics* (Christian Journals, 1982).

search is usually called metaphysics, but if one believes, as I believe, that the reality we experience is held in being by the will of God, then metaphysics become theology in its second role, sometimes called fundamental theology. If God is the ground of all that is, then theology in this mode will have to speak about all that is. It will be seeking to act as the great integrating discipline, taking the insights provided by the other forms of rational inquiry and aiming to set them within the most profound and comprehensive scheme of understanding available. I have written of theology in this role that

> If it is to lay claim to its medieval title of the Queen of the Sciences that will not be because it is in a position to prescribe the answers to the questions discussed by other disciplines. Rather it will be because it must avail itself of their answers in the conduct of its own inquiry, ... Theology's regal status lies in its commitment to seek the deepest possible level of understanding.[2]

On this view, those who are seeking understanding through and through—a natural instinct for the scientist—are seeking God, whether they name him or not. Bernard Lonegan put this with lapidary elegance: "God is the unrestricted act of understanding, the eternal rapture glimpsed in every Archimedean cry of Eureka."[3]

There is, of course, no simple procedure for constructing a metaphysics going beyond (meta) science. The tests for the validity of the endeavour will be comprehensiveness and coherence. Every new insight that the separate inquiries can provide will be of value in that task. I believe that the dynamical theory of chaos can afford us significant help. However, we shall need a certain intellectual boldness in the way we proceed. Metaphysicians and theologians, like cosmologists, face problems which cannot be tackled without a certain willingness to speculate. A delicate balance is required between enterprise and caution, neither fearing to stick out our intellectual

---

[2] Polkinghorne: *Science and Creation* (SPCK, 1988; New Science Library, 1989), p. 1.

[3] B. Lonergan: *Insight* (Longman, 1957), p. 684.

necks a bit, nor doing so to the extent that they become disengaged from the sober body of reliable knowledge.

Let me at least start with my head drawn tightly in. The theory of chaos tells us that those tame systems on which we all cut our dynamical teeth—like the steadily ticking oscillator or the ceaselessly revolving single planet—are quite exceptional. I call them 'tame' because small uncertainties in our knowledge, or small disturbances in circumstances, only produce correspondingly limited consequences for their behaviour. To all intents and purposes we can know what they are up to. Yet most of the physical world, even when we describe it in the apparently dependable terms of classical Newtonian dynamics, is not like that. In Popper's famous phrase,[4] there are many more clouds than clocks around. Once systems attain an even modest degree of complexity they become subject to an exquisite degree of sensitivity to circumstance that makes them intrinsically unpredictable. Their behaviour exhibits apparent haphazardness, but not to an unrestricted degree—which is why 'the theory of chaos' is really such an unfortunately inept name. Rather one sees a kind of structured randomness, as other speakers at this Conference have described for us. I think that what I have said so far is physics, and so, generally accepted. If I want to make metaphysical (and so, potential theological) use of it, I shall have to take the risk of moving on to more contentious questions.

The surprising feature of classical chaos is that it presents us with apparently random behaviour arising from solutions to deterministic equations. It is an oxymoronic sort of subject. The metaphysical question is, which shall we take the more seriously, the randomness or the determinism? Let me put it this way. The most obvious thing to say about chaotic systems is that they are intrinsically unpredictable. Their exquisite sensitivity means that we can never know enough to be able to predict with any long-term reliability how they will behave. Unpredictability is an epistomological statement about what we can know. In my metaphysical speculation I shall want to go beyond that to make an ontological assertion about what is actually the case for the physical world. I want to say that it is open in its

---

[4] K. Popper: *Objective Knowledge* (Oxford University Press, 1972), ch. 6.

process, that the future is not just a tautologous spelling-out of what was already implicit in the past, but there is genuine novelty, genuine <u>becoming</u>, in the history of the universe. How can I seek to justify taking such a possible but unforced step? I do so on two grounds:

1. The first is philosophical. Like most scientists I am a critical realist; I believe that our investigations lead us to a verisimilitudinous grasp of what the world is like. I say 'verisimilitudinous' rather than 'true' because our understanding is never absolute but may well require correction when we explore new regimes of physical phenomena. We possess maps of the physical world sufficiently accurate for many, but not every, circumstance. Critical realism is, of course, a contended position in the philosophy of science, but I believe it to be the only adequate account of what scientists are actually doing. I shall not argue the case here, having done so elsewhere,[5] but I shall be content to nail my colours to the mast.

    If you are a critical realist you believe that what we know and what is the case are closely connected. That assertion is really a working definition of what realism means. We critical realists have 'Epistemology models Ontology' written on our T-shirts. You can see how natural that is for a scientist by recalling the early history of quantum theory. When Heisenberg wrote his celebrated paper on the uncertainty principle, he was concerned with analysing what could be measured. His work was epistemological. It was not very long, however, before almost all physicists gave it an ontological interpretation. The main-stream understanding of quantum theory sees the uncertainty principle as expressing a genuine ontological indeterminacy, rather than a merely epistemological ignorance. In an exactly similar way, it seems natural to me to interpret the undoubted unpredictability exhibited by chaotic systems as pointing to a genuine openness in the process of the physical world. I shall say some more later about exactly how I interpret this.

---

[5] Polkinghorne: *Science and Providence* (SPCK/New Science Library, 1989).

2. The second ground for embracing openness lies in its providing the possibility of reconciling physics with our basic experience as human beings of responsibility and agency in facing an open future, which we play our part in bringing about. I know, of course, that the assertion of human free-will has often been questioned or denied by philosophers (what has not?) but once again I nail my colours to the mast. Ultimately, the denial of human freedom seems to me to be incoherent for, on its own terms, the assertion would then be the merely meaningless mouthing of an automaton.

I am seized of the necessity, so often expressed by Professor Prigogine,[6] of science's being able to describe a world of which we can conceive ourselves as inhabitants. It seems to me that an ontological interpretation of chaotic dynamics would be very helpful to that end. Let me emphasize that, if that is so, the gain is a gain <u>for physics</u>. I am not suggesting for a moment that until chaotic dynamics came along we were uncertain whether we possessed powers of choice and agency. We have always known that we do, a knowledge quite as foundational as any knowledge on which science is based. Indeed the whole scientific enterprise as pursued by scientists would not make sense without it, for if we were mere machines, we could not also be rational beings. The advance lies in seeing how that experience can be integrated in a promising way with what physics has to say.

These considerations encourage me to take the possible but speculative step of giving a primacy to behaviour over equations, of interpreting deterministic chaos as pointing to an actual physical world of subtle and supple character whose process is open to the future. On this view the deterministic equations from which our mathematical exploration began would be regarded as approximations in *an emergent downward direction* to this more subtle and supple reality.

The phrase in italics requires some explanation. Let me first nail another set of colours to that rather cluttered mast. (A talk such as this cannot be given without presuppositions, which it must be my responsibility, at the

---

[6] I. Prigogine and I. Stengers: *Order Out of Chaos* (Heinemann, 1984).

very least, to acknowledge explicitly.) I am an anti-reductionist. That is to say, I believe that biology is more than physics writ large; it has its own concepts and categories which are not reducible to those of physics alone. Those of us who hold this view normally—and to my mind persuasively—defend it in terms of a theory of upward emergence, asserting that new properties emerge as one ascends the ladder of increasing complexity of organization.[7] I suspect that this is only part of the story. If one believes that quarks and gluons are not more fundamental than cells or human beings—in other words, if one espouses a kind of ontological egalitarianism which does not assign a uniquely fundamental role to elementary particle physics, an egalitarianism which our holistic inclinations would certainly encourage—then there may well be emergence in both directions as one traverses the ladder of complexity.[8] On this view classical deterministic equations could be conceived of as emergent approximations to a more supple physical reality, as one made the simplifying assumption of treating the system in question as if it were isolatable from the rest of what is going on. (We know, of course, that the exquisite sensitivity of chaotic systems, their vulnerability to the slightest external trigger, means that they are never truly isolatable.)

You might suggest that this particular metaphysical knot might more readily be cut by involving quantum theory. After all, these exquisitely sensitive systems soon depend for the form of their future behaviour on details more fine than Heisenberg will permit. Would it not be better to justify the openness of the future by employing this sensitive enmeshment of the everyday world with the indeterminate quantum world? I can see more attraction in the suggestion, but I hesitate for several reasons. One is the gut feeling that everyday openness should not have to depend on goings-on in the microworld. Even if that were mistaken, there are other grounds for caution. One centres on the unresolved interpretative problems of quantum

---

[7] Polkinghorne: *One World*, ch. 6; A. R. Peacocke: *Creation and the World of Science* (Oxford University Press, 1979), ch. 4.

[8] Polkinghorne: *Reality*, ch. 3.

theory,[9] particularly the so-called measurement problem, whose perplexities arise precisely from our unsureness of how to treat the interaction between the macroscopic and the microscopic. Another reason lies in the difficulties that have been found in exhibiting chaotic behaviour in relation to the Schrödinger equation. I cannot but think that there must be at least some quantum analogue of classical chaos, even if it takes a somewhat different form (fuzzy fractals?), but it would be necessary to see this settled before one ventured very far along the route of involving a quantum basis for openness.

So far I have argued that chaos theory presents us with the possibility of a metaphysically attractive option of openness, a causal grid which delineates an envelope of possibility (it is not the case that anything can happen but many things can), within which there remains room for manoeuvre. How that manoeuvre is executed will depend upon other organizing principles active in the situation, viewed holistically. It is a long speculative step from complex dynamical systems to even the biochemical dance of a single cell, and a yet longer step to dare to speak about mind-and-brain, but having embarked on my metaphysical voyage I wish to push the exploration as far as I can. I do so mindful of the warning uttered by Thomas Nagel that those who today venture to speak about such matters are indulging in 'pre-Socratic flailings around.'[10] Nevertheless, we have to do the best we can, and I think there is at least a hopeful direction in which to wave our arms.

None of the classical accounts of mind-and-brain seem to me to be plausible. I cannot accept materialism's account, that all is matter and mind is a mere epiphenomenal ripple on its surface ('what the brain does'), because there seems to me to be an unbridgeable gap between talk of the firing of neurons, however complex the patterns considered, and even the simplest mental experience of perceiving a patch of pink, let alone ratiocination. I cannot accept idealism's account, that the mental is the real and the physical world a construct thereof, because the stubborn facticity of that

---

[9] Polkinghorne: *The Quantum World* (Longman, 1984; Princeton University Press, 1985), ch. 6.

[10] T. Nagel: *The View From Nowhere* (Oxford University Press, 1986).

world, so familiar to the scientist as he or she seeks to explore it and is continually surprised by it, speaks of an independent reality, standing over against our minds. Both materialism and idealism are implausible in their oversimplifications. Yet Cartesian dualism, with its talk of the extended stuff of matter and the thinking stuff of mind, leaves the two in unsatisfactory isolation from each other, with no real clue to how they might interact. That they do so interact is made clear by our experiences of willed action and the effects of being hit on the head with a hammer.

If it is not matter nor mind nor mind-and-matter, what then is left to us? I have suggested previously that we might try to consider a complementary metaphysic of mind/matter,[11] what the philosophers would call a dual-aspect monism: there is one stuff, but encountered in contrasting regimes it gives rise to what we call the material or the mental. The aim is to treat these two polar extremes in as even-handed a way as possible. Each pole would have its own anchorage in the appropriate dimension of reality; we participate in both a physical world and a noetic world.

The adjective 'complementary' is invoking a celebrated aspect of quantum theory, to which Niels Bohr drew particular attention.[12] If such invocation is to go beyond a mere slogan, it must indicate how the complementary poles are related to each other. In suggesting how this might be, I wish to use quantum theory as an analogical guide. One of its celebrated dualities is that of wave and particle. Quantum field theory tells us how the trick is done. A wavelike state is one with an <u>indefinite</u> number of particles in it (a possibility quite foreign to classical physics but permitted in quantum physics because of the latter's ability to mix together (superpose) states which classically would be immiscible). The key to complementarity always seems to lie in some dimension of indefiniteness. This suggests that mind/matter might be reconciled by their being different poles of the world's stuff in greater or lesser states of flexible organization; matter is the emergent downward side, mind the emergent upward side, where the arrow points in the direction of an intrinsic openness.

---

[11] Polkinghorne: *Creation,* ch. 3.

[12] N. Bohr: *Atomic Physics and Human Knowledge* (Wiley, 1958).

Chaotic dynamics would represent the first primitive stirrings of openness as one mounted the ladder of complexity leading from matter to mind. A chaotic system faces a future of labyrinthine possibilities, which it will thread its way through according to the indiscernible effects of infinitesimal triggers, nudging it this way or that. In the metaphysical extrapolation I am making, which sees chaos theory as actually an approximation to a more supple reality, these triggers of vanishingly small energy input become non-energetic items of information input ('this way,' 'that way,' as bifurcating possibilities are negotiated). The way the envelope of possibility is actually traversed depends upon 'downward causation' by such information input, for whose operation it affords the necessary room for manoeuvre. In this way I believe one can begin to get a glimmer of how the mental decision of my will to lift my arm might effect the physical action of its lifting, through messages conveyed by nerves from the brain to the muscles involved. The openness that chaos theory suggests provides the opportunity for holistic activity at the mental pole of that most complex and reflexively interconnected of all systems of which we are aware, the human brain. In a speculative way we begin to conceive a world of which we can indeed recognize ourselves as being inhabitants, made of one stuff which in appropriate regimes exhibits the behaviours we traditionally call material or mental, a world capable of sustaining causation by both energy and information, assigning instrumentality to both mind and matter.

Such a picture views human beings as psychosomatic unities—animated bodies rather than incarnated souls, to use a celebrated phrase. This is the prevailing (if perhaps not unanimous) tone of biblical anthropology (so that the Christian hope is not of survival but of death-and-resurrection). It is also fully consistent with our evolutionary insight that *homo sapiens* has arisen by continuous development from lower forms of animal life, and ultimately from the inanimate amino-acid-rich shallow seas of early Earth. Humanity's mental and spiritual capacities are then seen as the most strikingly realised potentialities of the world's stuff in open and flexible organization. The soul is not an extra spiritual ingredient injected at some stage into the body, and in principle separable from it, bur rather it is that holistic, almost infinite information-bearing pattern, carried by the body and maintained as the locus of our personal identity through all the unending changes of the atoms actually comprising our bodies at any particular instant of time.

Theologians have often, and rightly, wished to speak of God's continuing interaction with the world. Some have denied that we can say anything meaningful about its mode of operation.[13] I find this too intellectually despairing an attitude to take. Others (the process theologians) have wanted to use language of 'lure' or persuasion, based on a panpsychic view of reality.[14] I find this latter view totally implausible. Many others use words like 'influence,' 'guidance,' 'improvisation,' without indicating clearly what they could actually mean. A common thread in much of this discourse is the notion that in some way the analogical clue to how to speak about divine agency is provided by an appeal to the example of human agency. That is a strategy which I too would wish to adopt. The picture I have presented is of a world in which our willed action occurs through information input into flexible open process. That process is located in our bodies (presumably principally in our brains) but there is also open process elsewhere in the universe. I believe that God's continuing interaction with his creation will be in the form of information input into the flexibility of cosmic history. This is a theme which I have tried to pursue systematically in a recent book, *Science and Providence*.[15] Let me summarise some of the points which I think are relevant to the present lecture:

1. Though God is pictured as acting through open process, he is not pictured as soaking up all its freedom. It is an important insight of theology that in the act of creation, freedom is given to the whole cosmos to be and make itself. I have made this the basis of a free-process defence[16] in relation to physical evil (disease and disaster),

---

[13] A. Farrer: *Faith and Speculation* (A. & C. Black, 1967).

[14] See J. B. Cobb and D. R. Griffin: *Process Theology* (Westminster, 1976).

[15] Polkinghorne: *Science and Providence* (SPCK/NewScience Library, 1989).

[16] Ibid., ch. 5.

which parallels the free-will defence in relation to moral evil (the bad choices of human-kind). God neither wills the act of a murderer nor the incidence of a cancer, but he allows both to happen in a world which he has endowed with the ability to be itself. The balance between the room for manoeuvre which God has reserved to himself and that which he has given away is, of course, a delicate issue. It is also an issue familiar to theology, for it is the problem of grace and free-will, written cosmically large.

2. God's action within the cloudiness of unpredictable open process will always be hidden; it cannot be demonstrated by experiment, though it may be discerned by faith. Its nature also limits what it is sensible to pray for. Long ago, the Alexandrian theologian Origen recognized that one should not pray for the cool of spring in the heat of summer. Quite so: the succession of the seasons is a clockwork part of the physical world whose regular pattern reflects the faithfulness of the Creator, who will not overrule himself.

3. I have been criticized by some for what they believe is a return to the discredited notion of a God of the gaps. My answer would be that what was discreditable about that illusory deity was that the gaps were extrinsic, mere patches of current ignorance. As they disappeared with the advance of knowledge, the 'god' associated with them faded away as well. No one need regret his passing, for the true God is related to the whole of his creation, not just the puzzling bits of it. Yet if the physical world is really open, and top-down intentional causality operates within it, there must be 'gaps' (an envelope of possibility) in the bottom-up account to make room for this, whether I have identified their nature correctly or not. We are unashamedly 'people of the gaps' in this *intrinsic* sense and there is nothing unfitting in a God of the gaps in this sense either.[17]

---

[17] See ref. 8.

4. More subtly and disturbingly, one might fear that I have enmeshed God too closely with physical process, making the elementary and disastrous theological blunder of treating him as a cause among other causes, a mere invisible agent in cosmic process. I do not want to turn God into a demiurge and I do not believe that I have done so. His interaction is not energetic but informational. I believe my account is a kind of demythologization of what is unclearly articulated by those who use words like 'lure' or 'influence' or 'improvisation.'

5. The picture of physical process encouraged by the metaphysical exploration of chaos theory is one of a world of true becoming. I think that this has important implications for God's relation to time.[18] The picture of classical theism is that the eternal God sees all that happens in time *at once*. He does not foreknow the future; he knows it. It was a picture similar to that of classical relativity, with its space-time diagrams presented for inspection as frozen chunks of history. That is all very well for a deterministic world, which is really a world of being and not becoming. Chaos theory's intimation of openness changes all that. It seems to me that the eternal viewpoint is no longer a coherent possibility. The future is not up there waiting for us to arrive; we make it as we go along.

This has two consequences for our view of God. Firstly, he will have an intimate connection with the reality of time. That will scarcely come as a surprise to those acquainted with the biblical God of Abraham, Isaac and Jacob, involved in the history of his people. Of course God must also have an eternal aspect to his nature, for he is not in thrall to the flux of becoming, only in intimate and interacting relationship with it. One of the emphases of much twentieth century theology has been to seek to give an account of such

---

[18] Op. cit., ch. 7.

a dipolar (time/eternity) theism. I find the writings of Jürgen Moltmann[19] and Keith Ward[20] the most helpful in this respect.

The second consequence is more contentious, but it seems to me to be necessary. God does not know the future. That is no imperfection in the divine nature, for the future is not yet there to be known. Of course God is ready for it—he will not be caught out unprepared—but even he does not know beforehand what the outcome of a free process or a free action will be.

At first sight, this is a rather unpalatable conclusion for believers. We have become used to the notion that God's act of creation involves a kenosis (emptying) of divine omnipotence, as he allows something other than himself to be with the freedom with which he has endowed it. I am suggesting that we need to go further and recognize that the act of creating the freely other involves also a kenosis of the divine omniscience. God continues to know all that can be known—he possesses what philosophers call a current omniscience—but he does not possess an absolute omniscience, for he allows the future to be truly open.

We have come a long way from the analysis of logistic maps, or even the baroque splendours of the Mandelbrot set, to talk about the divine nature. This lecture has been a continuation of what is a perpetual preoccupation for me. I stand before you as a physicist and a priest, someone who wishes to take with equal seriousness what science has to tell us and what religion has to tell us. Of course their mutual interaction is sometimes perplexing, but I believe that it is often fruitful. There has been a long tradition that theology should be in dialogue with contemporary culture—whether it be Augustine with neo-Platonism, or Aquinas with Aristotelianism, or the puritan founding Fellows of the Royal Society with the emerging physics of their day. I think that the 'third revolution' of chaos theory affords promising opportunities for the continuation of that conversation.

---

[19] J. Moltmann: *God in Creation* (SCM Press, 1985).

[20] K. Ward: *Rational Theology and the Creativity of God* (Basil Blackwell, 1982).

CHAOS AND BEYOND
James Gleick

One of the dangers in a conference like this is that after a while you might start to think that it all makes sense. You listen to the eloquent presentations we've heard over the past two days and you begin to glimpse an intellectual framework with structure, logic, beauty. You feel it's only natural that Mitchell Feigenbaum found universal laws governing systems that make a transition from order to chaos. You feel it makes perfect sense that Benoit Mandelbrot spent the better part of a lifetime uncovering the strange patterns that exist in the paths of lightning bolts and the clustering of galaxies. You think, what else would someone like Stephen Smale do but invent a new approach to the mathematics of dynamical systems? What else would Heinz-Otto Peitgen do but explore the wonders of computer-generated fractals?

Unfortunately, what you've witnessed here is a little like what happens after a crime, when the perpetrators have had time to get their stories together. Sure, it makes sense now. Now chaos is a recognized movement in science. It's trendy. There are not only chaos conferences but chaos institutes and chaos journals. If you're a scientist for an oil company and you tell your boss—or your funding officer—that you're working on fractals, nowadays he thinks that's great. But it wasn't always that way.

We tend to suffer from the false impression that the progress of science is tidy, rational, somehow inevitable—that somewhere there is a list of problems to be solved and that scientists pluck the next one off the top of the pile. We tend to assume that ideas are communicated more or less smoothly from one place to another—that when a scientist discovers something worthwhile, the discovery finds its way to the other scientists who need to know about it.

What I've learned from the short recent history of this business called chaos is that none of these assumptions are true. And they are least true when scientists are producing something truly new.

The scientists who have come together to address this symposium were all, each in a different way, pioneers. They all experienced the difficulties of working on problems that their peers did not perceive as legitimate. They all experienced the awkwardness of trying to communicate discoveries that lay outside the ordinary way of thinking about things, discoveries that didn't have the decency to observe the customary boundaries between scientific disciplines. Chaotic dynamics and fractal geometry weren't on anyone's list of unsolved problems. They simply weren't a natural subject of study until these scientists made them one.

There's something about the ordinary way of portraying the history of science that works to clean up messes, like the street sweeper coming up behind the parade. In reality, the most creative science involves mistakes, guesses, false steps, miscommunication, uncertainty, a sort of stumbling through the fog that makes the exhilaration of true discovery all the more remarkable.

The ideas that come under the rubric of chaos, in particular, run counter to a conventional way of thinking about things. It's not always easy even for scientists to see that. All of us have preconceptions about the way the world works, the way nature works. These preconceptions, these biases, can be hard to shed.

I want to give you a simple example. Imagine a river. Not a real river, you know, the surface rippling in the squall, trout shivering under submerged rocks, currents carving the banks. That's not what I'm talking about. Nothing so fancy. Just imagine a river's basic shape, the way you might draw it on a piece of mental scratch paper.

You heard Benoit Mandelbrot say yesterday, "a river is not a straight line." But I'm willing to bet that virtually everyone here, even those of you who thought you were paying attention to Mandelbrot, has imagined a line of some kind. Maybe in your minds you drew it with some sort of curve or wiggle. That's just the way people think a river is shaped. When we speak of a river by name, we think of an entity flowing from one place to another—from source to outlet; from a mountain spring down to the sea. A geographer might ask, for example, what is North America's longest river? Is it the Mississippi—the river whose source is a little north of here? Or, is it the combination that doesn't quite have a name, the Mississippi/Missouri?

In reality the question makes no sense. In reality a river's basic shape—and it does have a basic shape, repeated wherever nature empties the land of water—is not a line but a tree. A river is fundamentally, in its very soul, a thing that branches. So are most plants—trees themselves, of course, bushes, ferns like the ones you saw today in computer simulation. So is the human lung, a tree of ever smaller tubes—bronchi, bronchia, bronchioles—intertwining with another tree, the network of blood vessels, the circulatory system.

So North America's longest river is actually a messy object spanning 31 American states and two Canadian provinces. It embodies, without distinction, the great tributaries that we think of as separate rivers—the Mississippi, Missouri, Ohio, Tennessee, Arkansas, Minnesota, and so on. Except in human perception and human language, nothing really separates its few wide and deep stretches from its many small and narrow ones. Although it flows inward toward its trunk, in geological time it grew and continues to grow outward, like an organism, from its ocean outlet to its many headwaters.

In the vernacular of this new science, it is fractal, its structure echoing itself on all scales, from river to stream, to brook, to creek, to rivulet, branches too small to name, too many to count. When a child draws a tree, green mass sits atop a brown trunk as if the basic shape were something like a popsicle. A child's cloud is a smoothly rounded bulk, perhaps with wavy or scalloped edges. These are not the clouds we see. They are highly stylized forms like the international symbols for Railroad Crossing or No Smoking. As children or as adults we own a repertoire of such stylized images. They're like ideograms in Chinese painting. They help us see. For without these templates, our minds are powerless to sift the welter of sensations that bombard us. But they hinder our seeing as well.

Before they were scientists, Mandelbrot and Feigenbaum and the other pioneers of chaos were people who saw nature's complexity a little more clearly than the rest of us. The rivers, the clouds, the snowflakes of our usual perceptual tool kits miss the point, the intricate recursion, the convoluted flows within flows within flows. Our mental lightning bolts are Z's, our volcanoes are inverted and decapitated cones, our rivers are lines. Nature's are not so simple.

This is not just a matter of terminology. It's a matter of true understanding, of recognizing and avoiding a kind of foolishness about nature. Again, think about the Mississippi River. How many tributaries does it have? How small is the smallest tributary? How long are all the branches added together? To make sense of such innocent-seeming questions, begins to require at least a feeling for the infinite. One hears such questions constantly, questions that cannot be answered as simply as people imagine. What is the average height of an ocean wave? What is the average size of icebergs breaking off the Antarctic Shelf? What's the average duration of the El Niño climate pattern or of an economic recession? When you hear questions like these, a new kind of bell should go off in your mind.

In the case of a river, the number of branches and their total length approaches infinity, just as the size of the smallest branch approaches the infinitesimal. And it has to be that way. What is a river anyway, but a thing that drains water from the earth's surface. In some sense, if it is to function that way, its fingers must penetrate every part of that area. Furthermore, a river is a dynamical thing. The microscopic rivulets that feed a river appear and disappear with the rains. At its farthest edges, a river's precise shape and size fluctuate dynamically with the day's rainfall or the year's wet and dry seasons, one brook appearing as another dries out.

Nor is there any simple dividing line between the transient and the permanent parts of a river. Fluctuations affect every time scale, from the seconds of an ephemeral rain shower to centuries and millenia. If you could film the life of a river and replay it in fast motion, you would see a symphony of rhythms, from the bass notes of the great branches to the treble flickering of the tips.

The traditional view of nature overlooks such rhythms. Many flows escape our perception because they're too slow or too grand in compass. A cloud floating overhead gives the illusion of a static object, simply borne from one place to the next by the wind. But that's not how it is. Even when a cloud looks most like frozen cotton candy, it slowly seethes and tumbles in the air. A motion picture film of clouds played 10 or 100 times too fast always shows the wildness that lurks in what seem like gentle puffs.

The contention of this new science of chaos—the motivating contention—is that the seeming irregularities in nature, even in rivers and clouds, can be contemplated, sorted, measured, and understood. Traditionally, science looked for a more conventional order in nature and treated the erratic as a side issue, an unpredictable and therefore unimportant kind of marginalia. Not any more. If any of you, for example, follow the technical literature of population biology, ecology, you may have noticed a striking change in attitude just in the last few years. For several generations, the dominating assumption in ecology was an attitude, a prejudice, that can be summed up in a single phrase: the balance of nature. When people go into a wild area and start tinkering, spraying insecticides, what is it they're disturbing? The balance of nature. The idea is that, left to itself, a wild ecosystem finds a harmonious balance and equilibrium. It settles down. It regulates itself. Of course, there's some truth to that idea. Complex, nonlinear systems, do become self-regulating. When suddenly perturbed, they can be thrown into regimes of behavior that are unexpectedly different. Witness, for example, the frightening appearance of ozone depleted holes over Antarctica in the past decade.

But ecologists, led by chaos theorists, have begun to realize that the balance-of-nature idea was profoundly naive. By and large, ecosystems may seek a sort of equilibrium, but that equilibrium bears as much resemblance to the traditional idea as a rollercoaster does to a golf cart. In their natural state, left all alone, ecosystems can fluctuate wildly, populations rising and falling, dying out and springing up again. Wildness really does mean chaos, not the slightly saccharine kind of harmony that ecologists used to imagine.

The same is true in economics. If you think the natural thing would be for a complex system made up of all the interacting, intertwining commercial activities of humanity to find some sort of static equilibrium, or to rise and fall in regular, predictable economic cycles, well, you're not alone. But, you're missing the boat. If you've listened carefully to the speakers over the past two days, you may well feel that a richer view of the possibilities is in order. Now, I'd like to offer another somewhat deeper example of a sort of perceptual prejudice that I believe has skewed the path of science—in this case fundamental physics—for much of this century. Consider the following statement:

At bottom, the root of unpredictability in the world is quantum uncertainty—the subatomic measurement uncertainty first quantified by Werner Heisenberg in the 1920s.

This is sort of a true/false test. True or false: the uncertainty principle of quantum mechanics is the fundamental cause of unpredictability in nature. I say unpredictability, but I might also say indeterminacy, or, in the evocative phrase that John Polkinghorne used, everyday openness. It's not an easy question, and of course, if you're smart you don't want to give a yes or no answer. You want to add a ten thousand word footnote. Nor is it easy to know what most physicists believe. But I think it's fair to say that many physicists have talked or acted as though they believe this is the case. Here's an example: "The uncertainty principle signaled an end to Laplace's dream of a theory of science, a model of the universe that would be completely deterministic. One certainly cannot predict future events exactly if one cannot even measure the present state of the universe precisely!"

That could have been any of a dozen prominent physicists but it happens to be Stephen Hawking in his recent book, *A Brief History of Time*. He adds: "Quantum mechanics, therefore, introduces an unavoidable element of unpredictability or randomness into science."

There's an inverse corollary to this statement, too. The corollary is, if it weren't for quantum uncertainty, the laws of elementary particle physics would allow one, at least in principle, to calculate and to predict everything. Hawking doesn't shrink from this implication. I'll quote again:

"Since the structure of molecules and their reactions with each other underlie all of chemistry and biology, quantum mechanics allows us in principle to predict nearly everything we see around us, within the limits set by the uncertainty principle."

You can see why the question starts to seem far from trivial. In practice, as Hawking goes on to say, you can't do the calculations. The equations are too complicated, and so on. But those are just the messy details. In principle, according to this point of view, the laws of elementary particles are the fundamental laws, and the ultimate limit on predictability comes from quantum uncertainty.

I don't happen to believe this. I think that as the lessons of chaos sink in, both versions of this statement are going to start to seem somewhat disreputable.

Are the fundamental laws of physics the laws that govern elementary particles? They are if you mean fundamental in a sort of geographical or anatomical way, to refer to the tiny things of which larger things are built up. But if by the fundamental laws you mean the laws with the greatest generality, the most profound laws, the laws with the greatest explanatory power, then I think you have to look elsewhere.

How do the laws of quarks and gluons or even the laws of quantum electrodynamics help us understand bigger things like the Great Red Spot of Jupiter, or the general problem of turbulence in fluids, or the formation of clouds and snowflakes, or the working of the human brain, or the balance of nature, or any other mystery of the universe as it exists outside of particle accelerators?

The truth is that they don't really. I would contend that if you could imagine a universe with no Heisenberg uncertainty principle—a universe in which the position and momentum of a subatomic particle could be specified simultaneously—you would have a universe in which it would be precisely as difficult as it is in our universe to predict next Sunday's weather; or to predict what will happen to the price of oil next month; or to predict just about anything about the behavior of any macroscopic complex system.

Much of the standard language of the physics community over the past two generations has embraced a reductionist perspective. At this conference, you heard John Polkinghorne declare that he was an antireductionist. Just a few years ago it would have been just about as likely for a high energy physicist to declare himself an antireductionist as to declare himself an Anglican priest. Physicists have believed and have stated proudly that before one can understand the whole, one must understand the parts.

Chaos is antireductionist. This new science makes a strong claim about the world, namely, that when it comes to the most interesting questions, questions about order and disorder, decay and creativity, pattern formation and life itself, the whole cannot be explained in terms of the parts. There are fundamental laws about complex systems, but they are new kinds of laws. They are mathematical laws of the kind that Smale, Feigenbaum, and Mandelbrot and others have discovered. They are laws of structure and organization and scale, and they simply vanish when you focus on the individual constituents of a complex system. I think we all understand that intuitively from our own everyday experience. I think that most of us now believe that when science comes to understand the interesting things about how the brain works, the things we all want to know—What is memory? What are thoughts? What are emotions? How are symbols stored and processed?—the answers to those questions are not going to depend on the particular chemistry of individual neurons. That chemistry may be what it is because of a thousand accidents in the history of life on earth. Intelligence is something else, something dynamical.

I think that the best quantum physicists have understood this all along. Having stressed what a great surprise chaos has been, now I would like to quote something Richard Feynman said 28 years ago:

> The complexities of things can so easily and dramatically escape the equations which describe them ... Man has often concluded that nothing short of God, not mere equations, is required to explain the complexities of the world.
>
> We have written the equations of water-flow. From experiment, we find a set of concepts and approximations to use to discuss the solution—vortex sheets, turbulent wakes, boundary layers. When we have similar equations in a less familiar situation...we try to solve the equations in a primitive, halting, and confused way to try to determine what new qualitative features may come out...Our equations for the sun, for example, as a ball of hydrogen gas, describe a sun without sunspots, without the rice-grain structure of the surface, without prominences, without coronas. Yet all of these are really in the equations; we just haven't found the way to get them out....
>
> The next great era of awakening of human intellect may well produce a method of understanding the qualitative content of equations. Today we cannot. Today we cannot see

that the water-flow equations contain such things as the barber pole structure of turbulence that one sees between rotating cylinders. Today we cannot see whether Schrödinger's equation contains frogs, musical composers, or morality—or whether it does not. We cannot say whether something beyond it, like God, is needed or not. And so we can all hold strong opinions either way.